artifact recovery

THE MATERIAL MANAGEMENT FIELD GUIDE

Nichole Doub

Artifact Recovery: The material management field guide
by Nichole Doub

ISBN: 978-1-957402-04-8

©2023 by The Society for Historical Archaeology

13017 Wisteria Drive, #395 Germantown, MD 20874, U.S.A. www.sha.org

Book Design by: Alan Barnett

All photos are courtesy of the MAC Lab (Maryland Archaeological Conservation Laboratory) except where otherwise noted.

Benjamin Ford, SHA Special Publication Editor

Table of Contents

Foreword

For much of the past fifteen years I've found myself in the simultaneously enviable and anxiety-inducing position of lab manager for a sizeable CRM firm conducting a very large, high-profile urban mitigation project in the eastern United States. To date, over two million artifacts have been recovered from this project, with the majority recovered from many hundreds of historic privy shaft features. The enormous scale of this assemblage is both exhilarating and intimidating, and it is in this unprecedented collection's management that I've worked closely with Nichole Doub. In *Artifact Recovery*, she makes clear that in any project, no amount of advance research can adequately predict everything that will be found. The archaeological record is always full of surprises, especially, it seems, when we are least prepared. At the start of my company's urban project, it was impossible to anticipate the tenacious durability of the archaeological record that we've since encountered in such a long-developed city, nor the rich diversity of the objects recovered. We never could have anticipated the leather fire helmet, the pressed-felt top hat, the rubber overshoes, the wooden smoking pipes, the tin toy soldiers, the cupreous candlesticks, or the 200-plus pound iron glassworks door, among many others. We have since developed near continuous consultation with Doub, and as both field and lab work continue on the project, so too does her input and work on its rare surprises.

While cultural contexts and artifact identification tend to be the focus of archaeological education, comparatively little emphasis has been placed on the actual materials that were used in the fabrication of the objects we recover or why that information might even be relevant. In *Artifact Recovery*, Doub successfully provides a basic introduction to the chemistry of those artifacts and materials most commonly recovered from North American post-contact sites, as well as context for how they deteriorate upon excavation and how our recovery methods, packaging, lab processing, and storage methods impact their preservation. The historic archaeological record can obviously yield a bewildering variety of objects, and in such an environment we all have blind spots. Doub acknowledges that no one can be an expert in every area of artifact identification, but she has created this highly accessible field and lab manual by reducing that record to the foundational material categories of metals, organics, skeletal materials, and synthetics.

In my lab, historic glass bottles have often been a material niche that staff have found challenging and difficult to master. I've made the most educational headway by adopting a similar strategy to that employed in this publication. Staff are taught to look past the many different shapes, sizes, colors, and types of decoration that make dating bottles so seemingly daunting. The ultimate goal is the realization that all bottles are made in relatively few ways, using only a limited number of mold types and finishing methods. By eliminating the extra "noise" inherent in a bottle's aesthetics, analysis becomes far simpler. In *Artifact Recovery*, Doub rarely discusses specific objects, even within figure captions. Her focus remains exclusively at the granular level, allowing readers to see through historical archaeology's diverse, "noisy" façade to understand that which is most important in artifact triage and conservation: the basic material an object is made of. Thus, while a fire helmet, shoe, bucket, horse harness, and flask cover may possess very different interpretive uses and values, the fact that all have been fabricated from leather means that their appropriate recovery methods in the field, packaging, lab processing, and storage methods are far more similar.

Beyond merely presenting the "how's" of artifact recovery and processing, Doub's explanation of artifact material chemistry makes plainly clear the "why's" as well. In my professional experience, perhaps the most fundamental aspect of successful management is ensuring that your lab staff or field crew fully understand the "why" in everything they do. In the absence of the "why," an archaeological lab becomes a mindless, automated production line, devoid of critical thinking. Mistakes are rampant and repeated in such an environment, productive collaboration tends to be lacking, and staff are frequently unable to grasp the gray areas that define so many corners of historical archaeology.

Many of the objects typically deemed worthy of formal conservation likewise often inhabit some of the less frequented corners of the archaeological material world, either due to rarity in the archaeological record itself or budgetary limitations preventing proper study and preservation. As a result, the majority tend to exist in most archaeologists' aforementioned blind spots. A recurring message throughout this entire publication is Doub's advocacy for professional collaboration. Consultation with a reliable and trustworthy conservator can serve to not only provide real-time support in the field and short-term advice in the lab, but can also aid in the identification of rarely encountered organic artifacts or those infrequently X-rayed rusty metal concretions. Even when recovered objects are identifiable from the start, further cleaning and conservation can often add an immense amount of valuable information.

My company's large urban project has proven challenging in that many of the properties we've investigated had relatively frequent resident turnover, in some instances with occupations spanning only a couple years, and in most cases lasting less than a decade. Such situations leave very little margin of error in our efforts to attribute artifacts and features to specific residents. Production dates derived from more commonly recovered and mass-produced artifacts like ceramic and glass vessels usually are not tight enough for us to achieve that desired level of clarity. On the contrary, there have been a number of instances where a particularly personal (as many are) conserved artifact has provided that missing link between the archaeological and historical contexts, be it through a uniform button, unique tool, or even merely a new TPQ. While we all have material blind spots, I highly recommend finding a knowledgeable, and well-connected, conservator to help "cover your back."

A resource like *Artifact Recovery* has long been needed in CRM, and I have every intent of making it required reading for my lab staff and proposing the same for our field techs, crew chiefs, and project managers. The information provided in this manual is only useful, however, if there exists a continuity of this knowledge across the field crew, lab staff, project management, and even client. In a CRM world where large firms are becoming increasingly more common, there can occur a disconnect between field and lab. It is only in the dissolution of this arbitrary divide, or at the very least in close consultation across it, that proper recovery, packaging, lab processing, and short-term storage can occur.

—Thomas J. Kutys
Archaeology Laboratory Manager, AECOM

Identification, Deterioration, and Preservation

Before the first shovel breaks the ground, archaeologists have extensively researched the site, pulled geographic information system (GIS) data, referenced historic maps and documents, and enlisted ground-scanning technologies. This research can anticipate the artifacts and materials that will be recovered, but excavation always has an element of surprise. Because of this uncertainty, it is impossible to be an expert in all areas of artifact identification. This chapter's goal is to provide guidance on the identification of material types, an understanding of how they degrade, and suggestions on the types of short- and long-term interventions that may be necessary for their preservation.

For most recovered artifacts, identification takes place in the field and in the lab using the observer's physical senses: sight, touch, smell, even hearing and taste. These observations can be supplemented with simple tools, such as magnification, and physical tests, such as observing interactions with light and magnetism. More advanced chemical tests performed in specialized laboratories will not be covered here as they are rarely applied save to answer very specific research queries.

Understanding how and why materials degrade helps us understand how to preserve them. Every material has a preferred state, a physical or chemical arrangement that uses the least amount of energy, which bonds it to other atoms or molecules to achieve equilibrium. Throughout history, people have learned how to manipulate these preferred states to form materials that have desirable properties that benefit their production or use, such as incorporating lime (CaO) to stabilize glass and increase its resistance to water, or adding fillers to clay to prevent ceramic shrinkage during firing. Some of these materials, such as ceramics, are relatively stable, only breaking down to physical forces through use, erosion, impact, etc. Some react to external forces in their environment such as heat, water, radiation, and electrochemical reactions. Organic materials are consumed by other organic organisms. Materials are constantly seeking to revert back to their natural state. Within the context of archaeological artifacts, the burial environment plays an integral role in objects' preservation and deterioration. Excavation from the burial environment immediately exposes artifacts to light and to changes in quantities of oxygen and water, removes physical support, and disrupts the state of equilibrium that enabled preservation.

Once a material has been identified and its state of preservation determined, it is possible to consider how that object can best be cared for. This may include passive stabilization (storing in a controlled environment), physical support (creating a mount or adding external materials to provide strength), or interventive conservation (treatments that address the physical and chemical effects of damage). Every object is unique, and while there are some generally accepted methods of preservation, there is no fixed recipe for conservation. Understanding potential options for treatment is useful in determining treatment priorities, setting a conservation budget, and effectively communicating with conservators and stakeholders.

Metals

Metals have a unique place in early material culture as their use and manipulation is one of the earliest industrial processes. Metals can be used individually or in alloys. Metals that were intentionally used in historic materials include gold, copper, silver, lead, tin, iron, mercury, and nickel (in chronological order by discovery/use). Aluminum wasn't widely produced until the late 19th century. The recovery and preservation of metal objects is very dependent on the burial environment. Acidic soils, burial environments with high levels of electrolytes (naturally occurring or from agricultural practices), high moisture contexts, and loose aerated soils tend to have a negative impact on metal preservation. Metals are best preserved in dry, compacted burial environments.

IRON

Historic iron artifacts, including cast iron, wrought iron, pig iron, and early steel, are all forms of iron combined with varying quantities of carbon that are then worked in different ways to form the final product. Iron alloys didn't come into widespread production until the mid-19th century, when the modern era of steelmaking introduced manganese, nickel, and chromium into the process.

Identifying iron on site may require multiple types of observation. Many iron artifacts will have some attraction to a magnet, but if the artifact is heavily corroded, the attraction may be very weak. Observing the corrosion product may provide more reliable identification. There are sixteen different types of iron oxides and oxyhydroxides that form the "rust" on iron. Their formation is dependent on several environmental factors in the burial environment. Most corrosion crusts are shades of brown, reddish-brown, orange, or yellow. The underlying surface of the iron objects may have a thin, stable layer of magnetite which appears as a dark blue-black patina. Weight may not be a good identifier for iron, as highly corroded iron is lighter than solid iron and the volume of the corrosion may conceal the size of the artifact contained within.

Iron recovered on site may be heavily obscured by corrosion product. It may be discerned by a perceived weight and some orange-ish coloration of the surrounding soil.

Detached fragments of iron spall are indicators of active corrosion.

Stable corrosion products are typically darker in color and compact. Active iron corrosion is generally a bright yellow or orange and is loose and powdery. Most iron artifacts recovered in the field will appear stable, but the introduction to new environmental factors post-excavation can initiate new corrosion events. In the lab, observations of small, detached corrosion fragments, spall, or depressions on the objects' surface with an orange center are indicators of active corrosion.

DISTINGUISHING CAST AND WROUGHT IRON

The two main distinctions between cast and wrought iron are their carbon content and method of manufacture. These give the artifacts produced very different material characteristics. Wrought iron has a very low carbon content (less than 0.1%) and is formed into an object through heating and working with tools. The structure of wrought iron is fibrous and corrodes along the inclusions in the metal, which gives it a rope-like appearance. Wrought iron has a higher tensile strength, making it appropriate for use in tools. Cast iron has a higher carbon content (between 2% and 4%), which is formed by melting and pouring the material into a mold. Cast iron is hard and brittle, making it inappropriate for tool use but ideal for objects, such as hollow projectiles that are intended to shatter, or objects where brittleness is not a factor, such as decorative railings or cooking vessels. Cast iron exhibits a "brittle fracture" that may have lines oriented perpendicular to the object's surface. Knowing the type of artifact and its use can help to differentiate between wrought and cast iron without additional analysis.

Left: The rope-like structure of degraded wrought iron is visible upon the removal of corrosion and in x-ray images.

Right: Cast iron has rough fracture edges.

DETERIORATION OF IRON

Iron ore is a mineral substance that is rich in iron oxides. Historic extraction of metallic iron depended on smelting processes that applied heat and a reducing agent to decompose the ore. The metal was then worked into the desired shape. This working process influences how iron deteriorates. Cast iron is produced by heating the metal until molten and then pouring it into a mold. The higher carbon content required to reach the molten state makes it more resistant to corrosion, if also more brittle. Wrought iron has been heated and then worked with tools. The working process concentrates the carbon and any impurities/slag into layers or strings, which give the material hardness. Wrought iron corrosion follows along the lines of inclusions which can give the degraded material a rope-like appearance.

During its period of use, iron artifacts often form a thin layer of corrosion on the surface called magnetite. This is an oxide of iron and one of the primary ores. As such, magnetite is very stable and can provide a protective layer against active corrosion prior to an object's deposition.

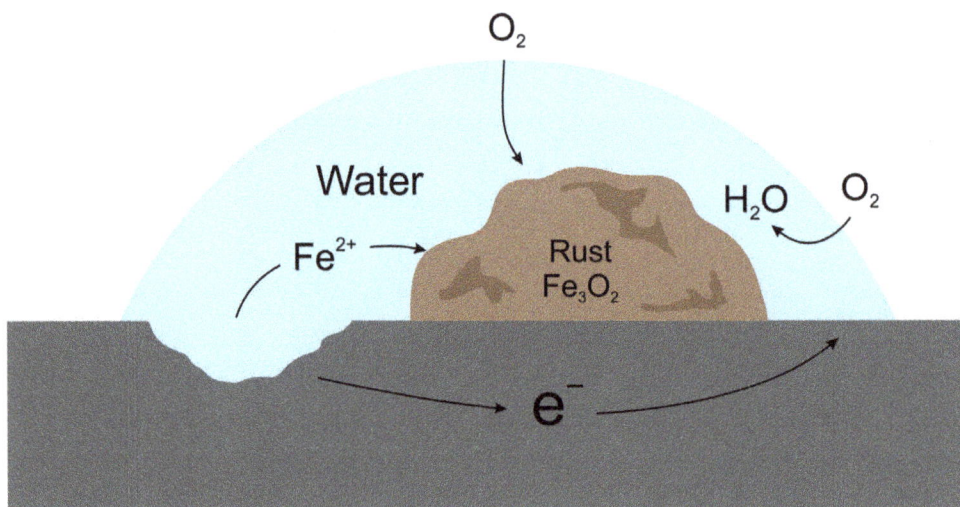

O_2

Water

Fe^{2+}

Rust
Fe_3O_2

H_2O O_2

e^-

Water interacts with oxygen to form a weak acid and electrolyte which dissolves some of the iron ($Fe2+$) and liberates electrons. As these electrons flow from the anode to the cathode site, the dissolved iron is converted to rust.

Iron corrodes in the presence of water and oxygen. It also requires an electrolyte to create the electrolytic reaction that converts iron back into an iron oxide (rust). These salts are present in varying quantities in both terrestrial and marine burial environments. As iron metal corrodes, the iron ions are drawn outward from the surface of the artifact and grow voluminous concretions by forming around the soil and particulates in the immediate burial environment, thereby obscuring the original surface of the object.

The oxidation of iron artifacts begins at the surface and works slowly into the material. Many iron objects may only have a thin metal core remaining, no metal at all, or at the most advanced stage of deterioration may be a hollow shell. In high chloride or marine environments, selective leaching of the iron can cause cast iron to become graphitized. The degree of deterioration is best observed with x-radiography.

Active corrosion commonly presents as a bright, powdery formation on the surface of an object, or in pits where spalling has occurred.

Conservation Terminology for Iron

Desalination

The desalination of iron uses caustic solutions that break down the bonds between the chloride ions and the iron to draw the soluble salts into the solution. The chloride concentration of this solution is monitored to chart changes in concentration until the chloride concentration no longer increases or a new wash extracts minimal chlorides. This does not ensure the complete removal of chlorides from the artifact, but it reduces the object's susceptibility for new outbreaks of corrosion and is used as part of a system for long-term preservation.

CORROSION INHIBITOR

A corrosion inhibitor is a chemical that reacts with the metal surface to form a stable passivating layer to provide protection to the metal core. Tannic acid is a common corrosion inhibitor used in the treatment of archaeological iron. When applied correctly to an iron surface, it forms a blue/black film of stable ferric tannate. This is a patina, part of the artifact, not a coating.

ELECTROLYTIC REDUCTION

Electrolytic reduction (ER) is a type of electrolysis using an electrolytic cell that converts electrical energy into chemical energy to drive a non-spontaneous redox reaction. The artifact serves as the positively-charged cathode; the negatively-charged chloride ions migrate away from the artifact into the solution or toward the anode. When excess energy is applied to the electrolytic cell, the water in the solution will decompose and produce hydrogen gas. This is generally an undesirable byproduct, but it can be used in certain conditions to break apart and loosen corrosion products. ER is used for desalinating iron and as an aid in corrosion removal. It is NOT a recommended technique for most archaeological artifacts as chemical reactions can be difficult to control and artifacts can be easily damaged. ER does not differentiate between the voluminous outer corrosion products and the dense corrosion layers that make up the object's original surface. Without close controls, it is easy to strip artifacts so that the surface detail is destroyed and only the core metal remains.

The example provided by the horseshoe illustrates the damage and loss of information that can occur when an artifact is subjected to electrolysis. Upon visual inspection the object is robust, but x-radiography shows a different story. The object is more heavily degraded than it initially appeared. The original surface of the object is suspended in the corrosion products and visible only in outline. The rope-like structure of the wrought iron is an indicator of the deterioration. While the metal core is present in some areas of the object, one quarter of the horseshoe has been completely replaced by corrosion product. Any treatment would be more damaging to the artifact than preserving the corrosion product, which is actually providing protection to the remaining metal.

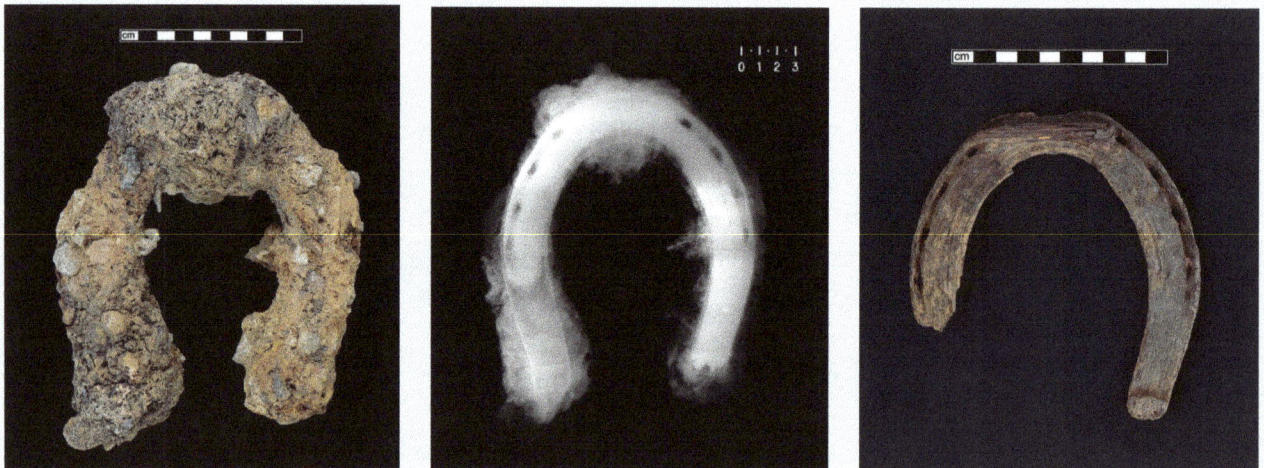

COPPER ALLOYS

Without analysis or detailed identification of an object, it is not possible to identify the specific alloy of copper in the field or processing lab. "Copper alloy" is a term that can encompass all historic combinations of copper, tin, zinc, lead, and nickel. The alloying of different metals gives different working properties and material characteristics. Bronze (copper + tin) is a ductile

Copper alloys are typically identified in the field by their characteristic green hue.

The break edge of a copper alloy object may reveal layers of different colored corrosion products, ranging from green to yellow to red, that have replaced the original metal.

alloy, can be worked or cast, and is much less brittle than cast iron. Brass (copper + zinc) is malleable and has a relatively low melting point for easy casting. Leaded copper and its alloys are easy to machine.

Copper alloys are most easily identified on site by their characteristic green- and blue-tinted corrosion products. Leaded copper may have spots of powdery white lead corrosion products. Well-preserved surfaces may have a thin, dark patina. A break edge that exposes the metal can vary in color from red to yellow to gold to brown depending on the composition of the alloy.

DISTINGUISHING BRONZE AND BRASS

There are some color differentiations between the alloys. Pure copper has a red hue. Brass is more yellow. Bronze ranges from a warm gold to brown depending on the amount of tin. Understanding the cultural context of a site and the use of an artifact may be a more accurate means of determining the composition of a copper alloy. However, detailed elemental analysis is often the only means to confirm composition. The color of the metal alone is an unreliable method of identification, and removing stable corrosion products to observe the metal may cause damage or start a new active corrosion event.

DETERIORATION OF COPPER ALLOYS

Copper metal is extracted through smelting. Since copper's melting point is within the same general range as other metals often mined from the same areas, historic copper is always an alloy. As control over the heating process became more refined, so too did control over the properties provided by these alloys. Tin is the most common alloy, followed by zinc and lead. Tin lends the alloy hardness, zinc creates a paler, more malleable metal, and lead improves casting properties. Copper itself is a nearly noble metal, meaning it is very resistant to corrosion.

During an object's use, a thin oxidized layer of cuprite will form on the surface. This layer is stable and often considered a desirable patina. However, both copper and its base alloying metals are susceptible to corrosion in the burial environment through oxidation, extrusion, and electrolysis. Copper corrosion crusts tend to be thin compared to the voluminous iron corrosion products. As with iron, chlorides facilitate the corrosion of copper by forming a porous corrosion product, nantokite (CuCl), also known as 'bronze disease,' that water and oxygen can penetrate to further the corrosion cycle.

Active copper corrosion forms pits in the body of the artifact, though some copper salts can migrate into the surrounding burial environment in the presence of moisture. These pits destroy the original surface of the artifact, whereas the stable corrosion layers can provide protection.

Stable surfaces on copper alloy artifacts have a wide range of patinas that can appear in shades of red, brown, blue, and green and are smooth and cohesive. Stable corrosion crusts are compact and may have a rough surface. Active corrosion generally appears as a loose, light green powder that can develop in isolated spots or across the surface of an object. Green staining of any paper products or labels in contact with the artifact is a clear indication of active corrosion.

Conservation Terminology for Copper Alloys

Desalination

The desalination of copper is often accompanied by a mildly alkaline solution that expedites the break down of bonds between the chloride ions and the copper to draw the soluble salts into solution. This solution also neutralizes the hydrochloric acid produced by the corrosion process and then reacts with the copper to produce a stable film of a copper carbonate (malachite). The chloride concentration of this solution is monitored to chart changes in concentration until the chloride concentration no longer increases or a new wash extracts minimal chlorides. This does not ensure the complete removal of chlorides from the artifact, but it reduces the object's susceptibility for new outbreaks of corrosion and is used as part of a system for long-term preservation.

Corrosion Inhibitor

A corrosion inhibitor is a chemical that reacts with the metal surface to form a stable passivating layer to provide protection to the underlying metal. Benzotriazole (BTA) is an effective corrosion inhibitor for copper alloys, forming an insoluble film that suppresses anodic corrosion.

WHITE METALS

White metals are a category that includes any pure metal or alloy that results in a light-colored, silvery metal. Given their similar appearance, white metals can be difficult to differentiate in the field. Corrosion products may provide clues to their identification.

Tin plating can be visually distinguished from other white metals by its luster.

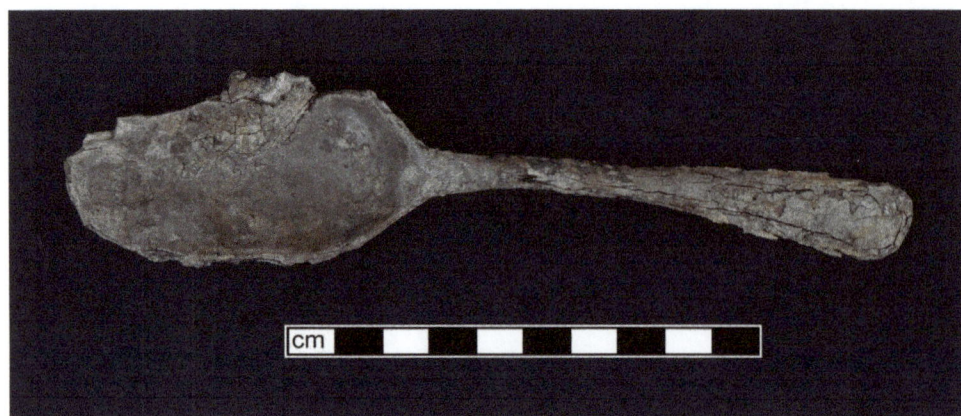

Pewter corrosion can be characterized by a cracked and/or warty surface often accompanied by a light brown crust overtop the grey metal.

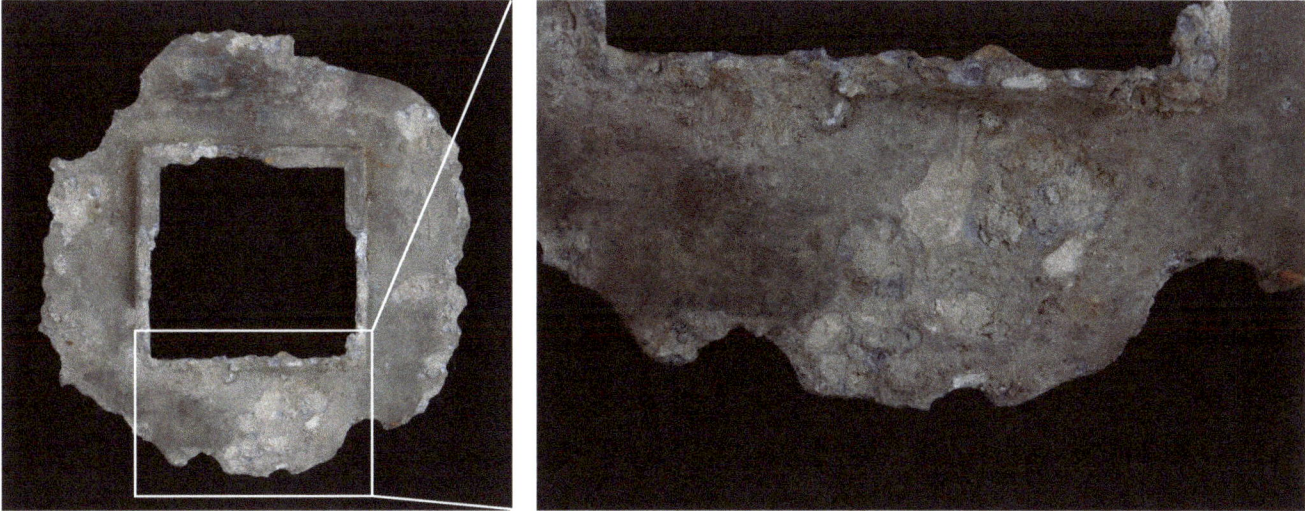

Britannia metal can be difficult to distinguish from pewter. Its corrosion often exhibits localized pitting as opposed to fracturing.

Stable lead corrosion products are a hard, dull gray and will often contain inclusions from the excavation soil. Lead artifacts that are actively corroding are bright white and powdery. Lead is also readily identified by its weight, as it is heavier than may be expected by the size of the object.

Tin is most often alloyed with lead to form pewter, or with copper and antimony to produce Britannia metal. Pewter has a distinctive warty or cracked corrosion crust, often with a discolored brown surface. Britannia metal is more lustrous than pewter and more resistant to corrosion, but it will suffer pitting. Both pewter and Britannia metal are used as a base for silver plating.

Aluminum was not in widespread use until the late 19th century. Aluminum is very lightweight compared to lead, although both have a similar silver-gray appearance. Stable aluminum corrosion forms as a hard, white crust; active corrosion is visible in the formation of blisters and pits.

Tin, zinc, silver, chromium, and nickel were all used as plating materials. Without elemental analysis it is difficult to determine the material. Instead, the use of the object is a better identification. Utensils, both iron and copper alloy, may have a silver plating. Copper alloy cooking vessels are more likely to have a tin plating.

DETERIORATION OF WHITE METALS

White metal is a combination of bright metal alloys that are often used for decorative purposes. They are primarily lead- or tin-based, and can also include zinc, silver, and nickel. The corrosion products often appear as warts or thin crusts that disfigure the surface of the artifact.

Lead is a corrosion-resistant material but is susceptible to deterioration in acidic environments. The most common corrosion product is lead carbonate ($PbCO_3$), which can provide a passivating layer of protection against further oxidation. However, depending on the conditions of the burial environment, and particularly the soil pH, lead can experience advanced deterioration. The corrosion replaces the original surface of the artifact and appears as a powdery white layer. As lead carbonate is soluble in acids, the corrosion can penetrate below the surface and cause pitting. Removal of the corrosion results in the removal of the object's surface layers.

Tin is another corrosion-resistant metal that is susceptible to attack by alkali and acid environments. Similar to lead, tin also forms a protective oxidized layer of corrosion on its surface.

Conservation Terminology for Lead

Electrolytic Consolidation

Electrolytic consolidation is a type of electrolysis that uses a low electric current to reduce a metal oxide (e.g., lead corrosion products) in a galvanic cell, converting the corrosion product back to stable lead. It should be noted that although electrolytic consolidation can be a very useful technique in preserving the original surface of a corroded lead object, it does alter the chemical structure of the lead, preventing future material analysis.

Before and after treatment of a lead bale seal using electrolytic consolidation.

COMPOSITE METAL ARTIFACTS

Some artifacts are made of more than one metal type to take best advantage of their physical properties. For example, a folding knife handle is made of a copper alloy because it is easily malleable and can be cast into the required

The less noble iron blade in this folding knife is corroding at an accelerated rate due to the contact with the copper alloy handle.

shape, but the knife blade is of wrought iron because it is more durable, stronger, and better able to keep an edge. In terms of preservation, the presence of two metals in close contact with one another can accelerate deterioration. Galvanic corrosion occurs when one metal corrodes preferentially when it comes in contact with another metal in the presence of an electrolyte. In these systems, the more noble metal will be preserved to the detriment of the less noble metal. Noble metals have a more stable electron configuration and are less prone to react to external forces that cause corrosion. In the case of iron and copper, the copper is more noble and will be better preserved than the iron, which will experience more advanced deterioration.

SILVER AND GOLD

While silver and gold can be discovered in their solid forms, it is more common to find objects that have a thin surface layer of these valuable metals. Gold presents in a range of colors, depending on its purity and alloyed materials, from bright yellow, to almost white, to pink. It is easy to distinguish in the field by color but can be obscured by the corrosion products of neighboring base metals. Silver presents from bright to dull gray, depending on the degree of oxidation. Similar to gold, it can be obscured by base corrosion products, most commonly copper alloy which, due to its mechanical strength, casting properties, and lesser cost, is used as a base for gilding and plating.

Zinc
Iron
Tin
Lead
Copper
Nickel
Silver
Gold

More Reactive

Less Reactive

Nobility scale of metals.

The iron corrosion has pushed away the thin layer of silver plating. This causes significant loss in the burial environment, often leaving only traces of the plating trapped in the corrosion.

SILVER AND GOLD DETERIORATION

Gold and silver are both noble metals that are resistant to chemical deterioration. In the context of archaeological preservation, this means that they survive very well with minimal degradation. Although this is true for objects that are cast of these materials, the majority of artifacts contain only small portions of gold or silver. Gold leaf and silver plating are more common and are subjected to material damage or loss resulting from the deterioration of their substrates. These thin layers can be easily lost in the bloom of iron or copper corrosion. While chemically stable, silver and gold are very soft and susceptible to mechanical damage.

Silver does not typically produce a voluminous corrosion product unless it is heavily alloyed or exposed to excess chlorides. Silver is often alloyed with copper, which can confuse the identification because of a green or reddish corrosion product. Silver may appear bright when initially excavated but it will tarnish quickly, turning a dull gray. Silver is susceptible to oxidation, also referred to as tarnish. While this may be aesthetically unappealing, it rarely causes any damage to the object. Because of the gray/black appearance of the tarnish, it may be difficult to distinguish between silver objects and white metal objects in the field. The use of the object may be a more useful identifier; silver and silver-plated objects are generally ornamental or have decorative features. As soft metals, silver and gold do not support robust utilitarian functions except as ornamentation over a stronger material, such as a copper alloy.

Conservation Terminology for Silver

Polish

Silver polishes are not recommended for use with archaeological recovered materials. Many metal polishes contain abrasive compounds that remove the tarnished surface from the object. Many commercially available polishes contain ammonia which can dissolve copper in the alloy or the base material under the silver plating.

Electrolytic Reduction (ER)

Similar to the process for desalinating iron, ER can be used to remove the tarnish from the surface of silver artifacts. By setting up an electrolytic cell, the tarnish (typically a silver sulfide) is reduced back to silver metal while the sulfur is drawn into the solution or bonds with the hydrogen gas produced in the reaction.

Organics

Organic material is less represented in the archaeological record because it requires a narrower range of environmental conditions for its preservation. And yet it makes up an equal or greater percentage of the material culture, from building materials to clothing to tools and more. Recording and preserving these materials is a high priority, both for the information they may represent as well as for their fugitive nature. Organic materials are best preserved in environments that are inhospitable to the micro-organisms that feed on organic materials, particularly dry or low oxygen burial environments.

WOOD

Waterlogged white oak viewed under magnification.

Wood has a distinctive grain pattern that is a byproduct of the cut angle in relation to the angle of the wood's growth rings. This pattern may be obscured by soiling or staining on the object's surface but will be visible under magnification. Hardwoods can be distinguished from softwoods by their pronounced growth rings. Wet and waterlogged woods often have a pulp-like texture and may exhibit some fibrous loss on the surface. Such wood is typically dark in color. Species identification is best carried out by a specialist, who may need to take a small sample for destructive analysis using thin section microscopy.

WOOD DETERIORATION

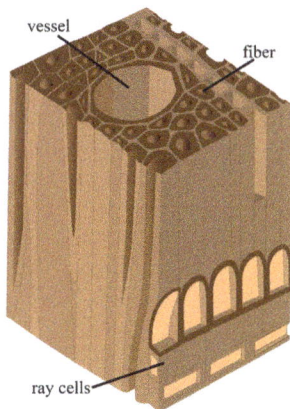

Hardwoods (such as oak) are made of vessel, fiber, and ray cells.

Wood is an organic material that is subject to deterioration by both biological and chemical attack. Wood is comprised of two major components, lignin and carbohydrates (cellulose and hemicellulose). The lignin is an organic polymer that makes up the structural support tissues within the plant, particularly the cell walls. It is resistant to degradation, unlike cellulose and hemicellulose, which are highly susceptible to biological attack from microscopic organisms (fungi, bacteria) as well as hydrolysis (the chemical breakdown due to water).

The survival of wood artifacts in the archaeological record is dependent on the conditions of the burial environment. Environments that reduce the activity of microorganisms, such as very dry or very low oxygen environments, have a higher probability for organic preservation.

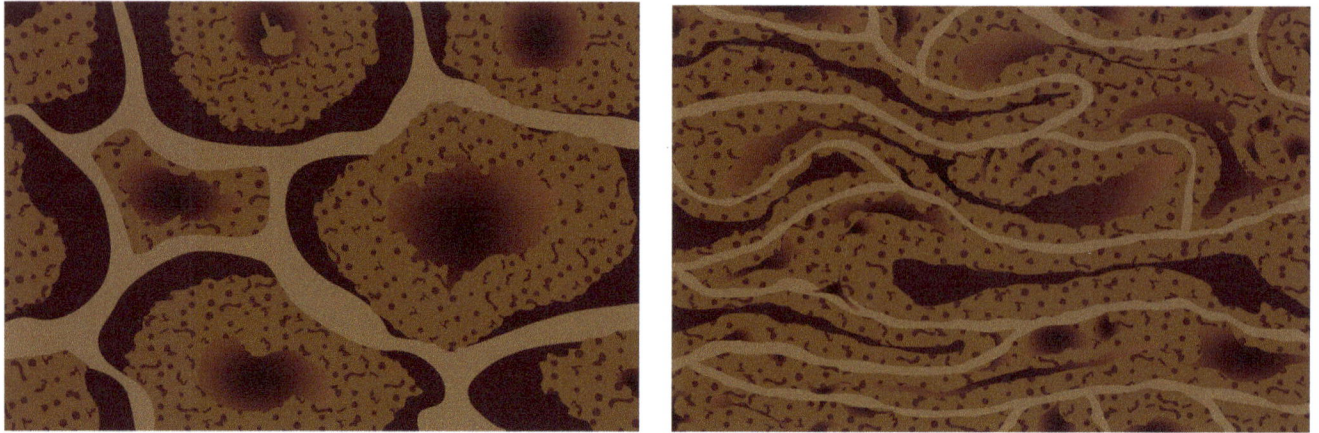

In the case of preserved wet and waterlogged wood, the object may have been buried under silt or clay, which created an anoxic environment that slowed the activities of anerobic bacteria. The surrounding water was then absorbed into the cellular structure, providing support to the degraded cell network. Cellulose and hemicellulose make up 60–80% of the wood's total mass. Without controlled drying and the introduction of a cellular bulking agent, the cells can collapse, resulting in cracking, shrinking, and warping of the artifact.

Jet is an organic mineraloid that is formed when wood becomes waterlogged and is then subjected to geological pressure and heat. Jet is very lightweight, warm to the touch, and a dull waxy black that can be polished to a high sheen. Its surface often has distinctive fissures and may contain the preserved cellular structure of its wood origins.

Charcoal is the carbon residue from wood that has been heated in a low-oxygen environment that burns off all volatile compounds. Charcoal can also be produced inadvertently by burning wood, as in cook fires. Charcoal is lightweight, black, and fractures along the natural channels occurring in the wood structure.

Left: degraded waterlogged wood cells where water is filling the spaces between the walls and lumen. **Right:** collapsed cells in the absence of water.

Hot Pin Test

The hot pin test is a simple material characterization test using heat to excite odor-causing molecules in a substance. Friction, rubbing an object between fingers, may produce enough heat to cause an odor, if the object is large and robust enough to withstand rough handling. For small, fragile objects, a heated metal pin or needle can be applied to a discrete surface. Depending on the chemical composition of an artifact, distinct odors will be produced and can be identified by a discerning observer. The hot pin test is considered a destructive testing method and should be used with care.

➤ **A hot pin test for jet will produce an oily odor.**

Jet is rich black or dark brown in color. When broken, the edges may have a similar shear to glass. Jet is opaque when held up to light.

Conservation Terminology for Wood

Consolidation

Dry wood artifacts may experience some deterioration to the outermost surface layers, while the core of the artifact remains robust. A consolidant, a dilute adhesive, can be introduced to add structural support and provide strength to weakened areas.

Freeze Drying

The greatest risk of damage to waterlogged wood is allowing the wood to dry. As water leaves the wood, the deteriorated cells that were relying on the water for support collapse. Bulking agents can be added to the wood to replace the support that was provided by the water. A variety of bulking agents have been used by conservators; the most widely used is polyethylene glycol (PEG). PEG is a popular conservation material as it is chemically inert, non-toxic, and water soluble. Once an artifact is fully impregnated with PEG, the water is removed. Natural drying methods are ineffective, as the PEG can migrate out of the wood with the water and collect on the surface, which leaves a dark and waxy appearance on the object. The wood can also be damaged by natural drying: the strong capillary action of water can act on the cell walls and collapse the cells as the water pulls on the surfaces while exiting the structure. This can cause warping and cracking of the artifact. The preferred method of water removal is vacuum freeze drying. The PEG-impregnated artifact is placed into the chamber of the freeze dryer, which creates an environment with very low temperatures and pressures at a point where the water in the artifact sublimes. Sublimation is the transition from a solid to a gas without passing through a liquid phase. By removing water in a gaseous phase, conservators avoid subjecting the weakened cells to the stronger forces of water in its liquid form. As the water sublimes, the PEG remains in the cells in a solid state, providing strength and support to the structure.

PEG impregnation treatment is irreversible and can impact future analysis of the artifact. All analysis should take place prior to conservation treatment, or an untreated wood sample should be retained for future researchers.

1. Waterlogged artifact, cells contain both free and bound water.
2. Polyethylene glycol absorbed into the cell structure. 3. Artifact frozen prior to drying.
4. In vacuum freeze dryer, the water sublimes, leaving the PEG to provide support to the degraded wood structure.

Solvent Drying

Small wet wood artifacts may also have the option of solvent drying. Where PEG and freeze drying are not available or are cost prohibitive, this technique can remove water from waterlogged wood without the adverse effects of natural drying. To avoid subjecting the wood to the strong capillary forces of water, the liquid water is slowly replaced through many exchanges in baths of less polar solvents with higher evaporation rates. The artifact can then be slowly dried. This method is not appropriate for larger artifacts as it is difficult to control the drying process, expensive, and poses health and safety risks due to the high volume of solvents required.

LEATHER

Leather can appear in a wide range of preservation states. Dry leather may be hard and brittle or soft and flexible. Wet leather may be robust and pliable or spongy and crumbling. Most archaeological leather is darkened due to oxidation of fats and oils in the leather and is prone to staining from organic matter and neighboring metal artifacts in the burial environment. Dry leather in particular may have fine cracking over the surface. Leather can be best identified by the grain pattern on the hair side of the material or the fibrous texture along the flesh side or any torn edges.

Left to right: Waterlogged leather viewed under magnification. Surface grain, flesh side, and fibrous torn edge.

LEATHER DETERIORATION

The traditional definition of leather is animal skin that has been processed to remove the epidermal and fatty layers and then tanned to stabilize the collagen. Collagen is the main structural protein in skin and gut, which can be broken down via hydrolysis, ultraviolet (UV) radiation, microbial decay, and the oxidation of the amino acids that hold the collagen network together. Archaeological leather is a broad category of artifacts made from animal skin and gut, including fur hides, rawhide, vellum, parchment, gut, and sinew. Many factors go into the preservation of archaeological leather artifacts including the type of animal skin, part of the animal used, method of tanning/preparation, and burial context.

As with all organic materials, archaeological leather survives in burial environments that are hostile to biological activity. Archaeological leather can be found in dry climates, but it is often rigid and brittle. Once leather has lost the water content needed to impart flexibility and movement in the collagen fibers, it will harden and embrittle as the fibers shrink irreversibly. Leather artifacts preserved in wet, anoxic environments have the benefit of retaining some flexibility.

Leather

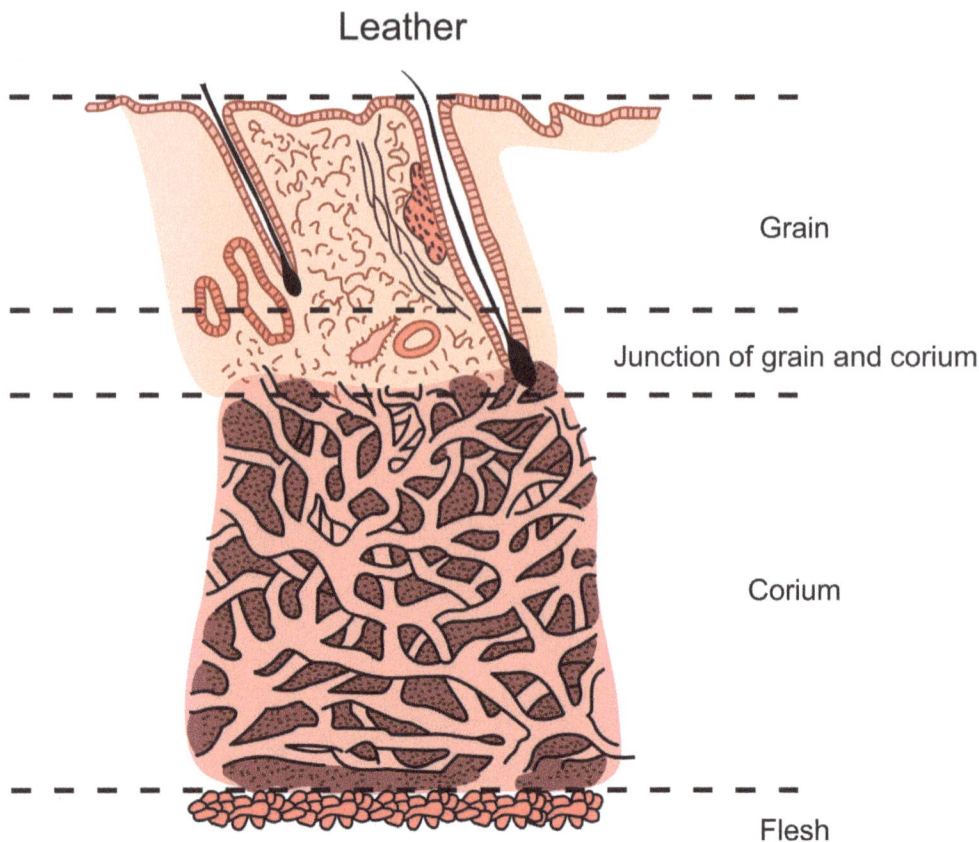

Grain

Junction of grain and corium

Corium

Flesh

The different layers of skin have unique physical properties that lend the treated leather to certain uses. For example, top grain is more durable and robust, while the corium is softer and more flexible. These properties impact the likelihood of preservation.

Conservation Terminology for Leather

Humectants

Wet and waterlogged leather can be treated by impregnating the leather with a humectant, typically polyethylene glycol or glycerin. Humectants attract and retain water, lubricating the fibers, which allows the dried leather to retain shape and flexibility. Freeze drying is the most effective method for the controlled drying of wet leather.

RUBBER

Natural rubber is the latex of the rubber tree and is thermoplastic (molten when heated and hard upon cooling). Vulcanized rubber is produced when natural rubber is heated with sulphur to make a harder and more elastic compound which remains hard and rigid when heated.

Vulcanized rubber can have many of the same visual and physical characteristics of leather, including a dark color, flexibility or rigidity, and fine cracking. Natural rubber will be light gray or off-white in color. It may be possible to distinguish the rubber by observing mold seams or textile backings or impressions.

Left: Many rubber objects have visible seams from their casting molds, as in this rubber button.

Right: Degraded rubber artifacts have a pattern of cracking deceptively similar to leather. (Photo courtesy of AECOM, PennDOT, and FHWA I-95/Gerard Avenue Interchange Improvement Project)

RUBBER DETERIORATION

Vulcanized rubber contains sulfur compounds that react with water to produce sulfuric acid, which both accelerates its deterioration and can affect other neighboring artifacts.

Natural rubber is prone to oxidation and can become cracked and brittle.

➤ **A hot pin test will produce smoke and the smell of burning rubber or sulfur.**

Conservation Terminology for Organics

Microbial Growth

Organic materials, particularly wet artifacts or those stored in high-humidity environments, are susceptible to biological attack. Mold growth can be a localized issue; it can also consume the whole artifact as well as its packing materials. Conservators may add biocides during treatment to mitigate microbial growth, though this must be done with caution as it can have long-term safety implications for collections personnel. For short-term mitigation of an active infestation, wet artifacts can be gently washed with potable water while the lab technician is using appropriate PPE. The objects can then be sprayed with a 50/50 mixture of ethanol and water. Where ethanol is not available, denatured alcohol is an appropriate alternative. The alcohol will have a drying effect on the mold spores and slow the continued growth. For dry artifacts, the mold can be dry brushed into a HEPA-rated vacuum and then lightly sprayed with the 50/50 ethanol mixture. Be aware that solvents can negatively impact pigments, textiles, and certain forms of chemical analysis and should only be used on a case-by-case basis rather than as a blanket treatment. Mold exposure can have serious health risks; safety precautions (i.e., the use of gloves, masks, air extraction, and HEPA vacuum filters) must be observed.

Skeletal Materials

Skeletal materials are the rigid components of living organisms. For vertebrates this includes bone, antler, horn, ivory, baleen, and tortoise shell. Nonvertebrates produce a range of skeletal materials, but here we have singled out mollusk shells because of their abundant use in objects recovered from archaeological sites.

BONE

Bone is made up of two structures, the smooth, hard, outer cortical bone and the porous interior trabecular bone. The porous structure of the trabecular bone is a useful identifier; however, worked objects are made from the cortical or lamellar bone. Depending on staining from the burial environment or bleaching from UV exposure, bone can range in color from off-white to dark brown/black. Examination under magnification may reveal small pits, formed by blood vessels, on the surface. Because of these microscopic pores, bone will stick to your tongue where wood will not. This is a quick field test, but it may not be appropriate for all objects or excavation contexts.

Burning bone at high temperatures results in discoloration of the material. Charred bone appears black, whereas calcinated bone will turn blue-gray or white.

Blood vessels are visible on the worked surface of a bone toothbrush.

Burnt bone can take on a blue/gray hue or mottled white appearance depending on the degree of heat and oxidation.

BONE DETERIORATION

Bone is composed of collagen wrapped in a hard crystalline structure of calcium phosphate ($Ca_3(PO_4)_2$), which grows in layers as the animal increases in size. Soil pH and moisture have the greatest role in deterioration; these factors contribute to the leaching of calcium and phosphates from the bone, which causes weathering of the outer surface and a loss in rigidity. Aerobic environments can also accelerate deterioration, as microorganisms decomposing the collagen produce organic acids that can promote weathering of the bone.

Antler, when worked, can be very hard to distinguish from bone. Charred bone and antler can be easily mistaken for wood. Under magnification the nervous and vascular canals are more prevalent than in bone.

Antlers, as an extension of an animal's skull, are structures of bone and deteriorate in the same way.

➤ **A hot pin test produces a burnt hair smell.**

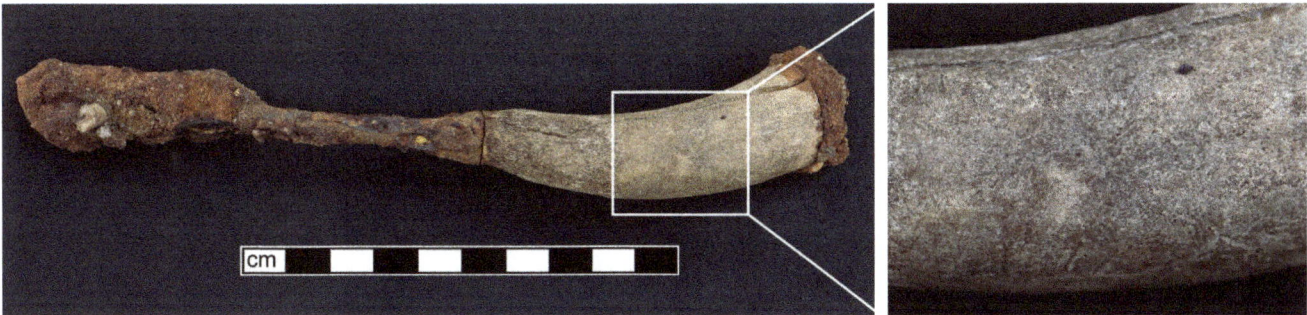

Antler is very porous and can be identified by micro vessels on the worked surfaces.

Conservation Terminology for Bone and Antler

Consolidation

Dry, friable bone and soft, wet bone artifacts can benefit from consolidation, the application of a dilute adhesive, when there is a high likelihood of material loss. The use of consolidant will depend on the presence of water and the amount of strength that needs to be imparted. Certain adhesives, such as polyvinyl acetate (PVA), are water soluble and will work with the water in wet bone but will not penetrate far beyond the surface of dry bone. This may result in the consolidated surface layers separating from the rest of the object. Other adhesives, such as Paraloid B-72, are not soluble in water. Paraloid B-72 suspended in a polar solvent (i.e., acetone) will easily wick into dry bone, but it will settle on the surface of a wet artifact and turn opaque as it reacts with the water.

Consolidation may be an irreversible treatment and can impact future analysis. All analysis should be carried out prior to treatment.

IVORY

Ivory can be difficult to identify by appearance alone. The characteristic striations, or Schreger Lines, can be difficult to distinguish from bone; one must look carefully at all the worked surfaces in order to find a cross section that reveals the pattern of micro-tubules unique to ivory. Elephant/mammoth ivory has a characteristic cross-hatched or stacked chevron pattern. Elephant is the most common ivory source recovered in the United States from historic contexts. The distinction between mammoth and elephant ivory requires careful examination and measurement of the patterning. Walrus ivory is characterized in cross section by a core with a marbled, walnut-like appearance. Celluloid and casein were popular in the production of imitation ivory. For additional information on identifying ivory sources, the World Wildlife Fund has published an update to their Identification Guide for Ivory and Ivory Substitutes (2020).

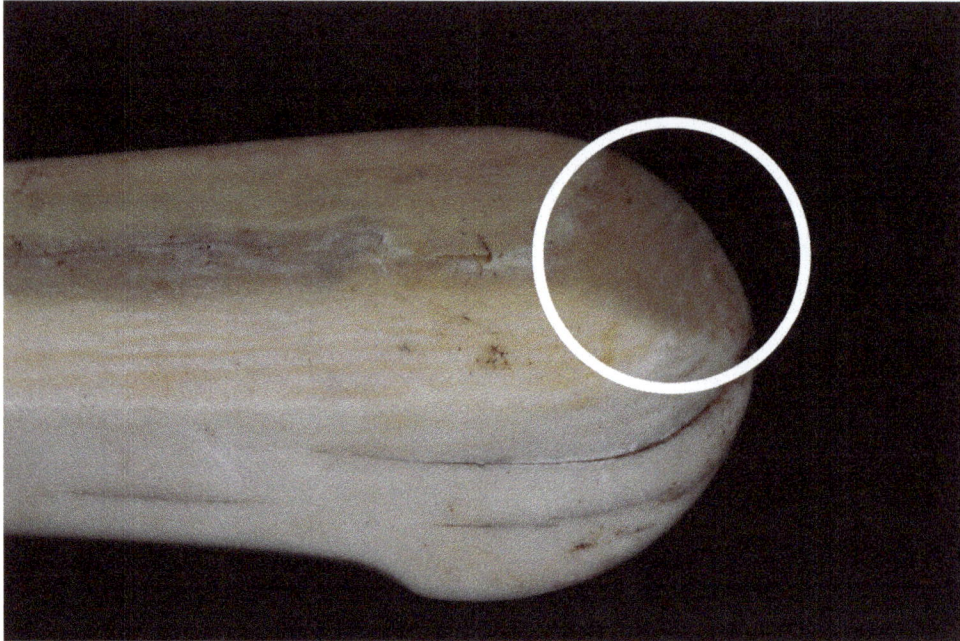

Longitudinal Schreger lines along the side of an ivory (elephant) utensil handle, with faint radial lines on the butt.

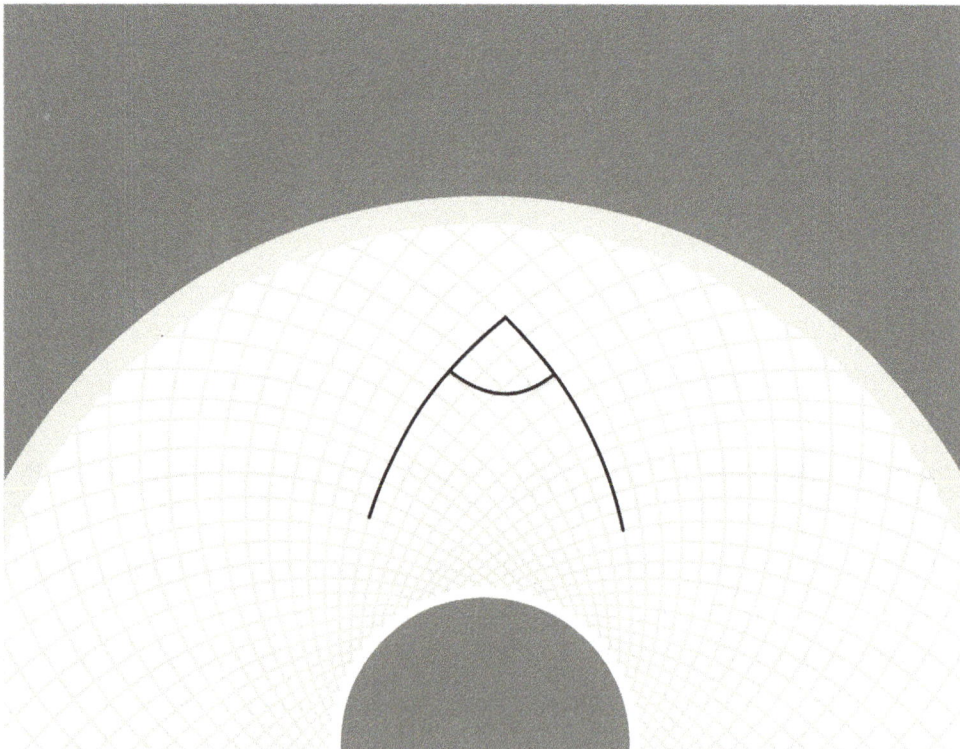

The angle of Schreger lines can be used to determine the species of elephant or mammoth ivory.

IVORY DETERIORATION

Ivory describes an object produced from any mammalian tooth or tusk. Teeth and tusks are made of the same physical structures: pulp cavity, dentine, cementum, and enamel. The pulp cavity is the empty space within the tooth/tusk. Dentine, the main component of ivory objects, forms a thick, consistent layer around the pulp cavity and makes up the majority of the

Ivory is very susceptible to moisture and changes in relative humidity, which can cause delamination and cracking.

tooth/tusk. The dentine contains a microscopic structure of tubules that radiate outward to the cementum layer that surrounds the tooth/tusk. These canals are unique configurations in different types of ivory. Enamel is the hard tissue that covers the surface of the tooth/tusk; it receives the most wear from use. Due to the multidirectional structure of ivory, it is very susceptible to warping and cracking, especially for thin artifacts. Chemical and bacterial attack degrade the protein components within the ivory, while acidic soil conditions impact the inorganic components, resulting in a soft and crumbling matrix.

➤ **Ivory is cold to the touch.**

➤ **Using long-wave (365 nm) ultraviolet light, ivory will fluoresce brightly, whereas most plastics and resins appear either dark or dull.**

➤ **Ivory is unaffected by the hot pin test.**

TORTOISE SHELL

Tortoise shell is a natural thermoplastic material that can be heated, fused, and pressed into many decorative shapes and thicknesses. Degraded tortoise shell may delaminate as the layers of keratin separate. Observation under magnification may reveal the layered structure as well as a spotted micropattern that makes up the dark patches. Casein, celluloid, and cellulose acetate were also used to produce imitation tortoise shell.

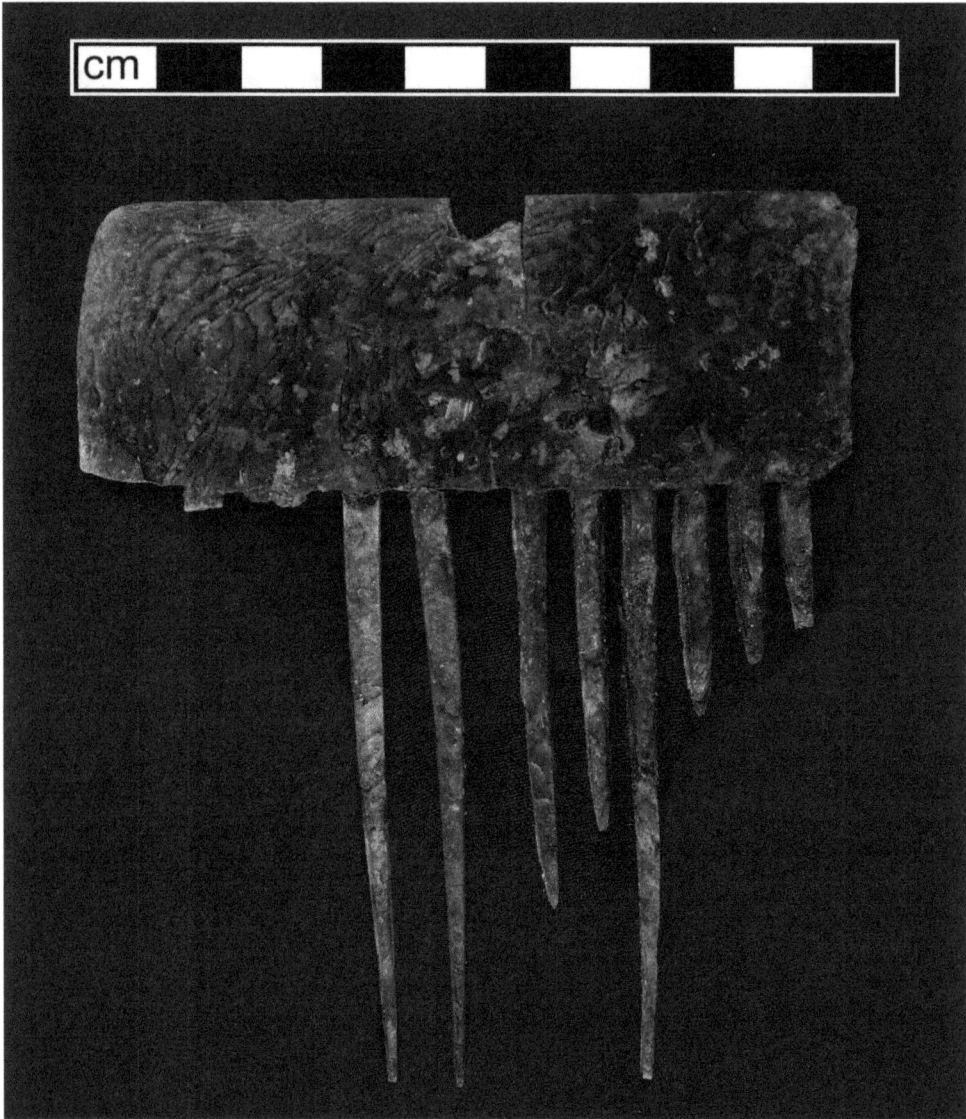

Tortoise shell comb showing distinctive mottling, layering, and delamination.

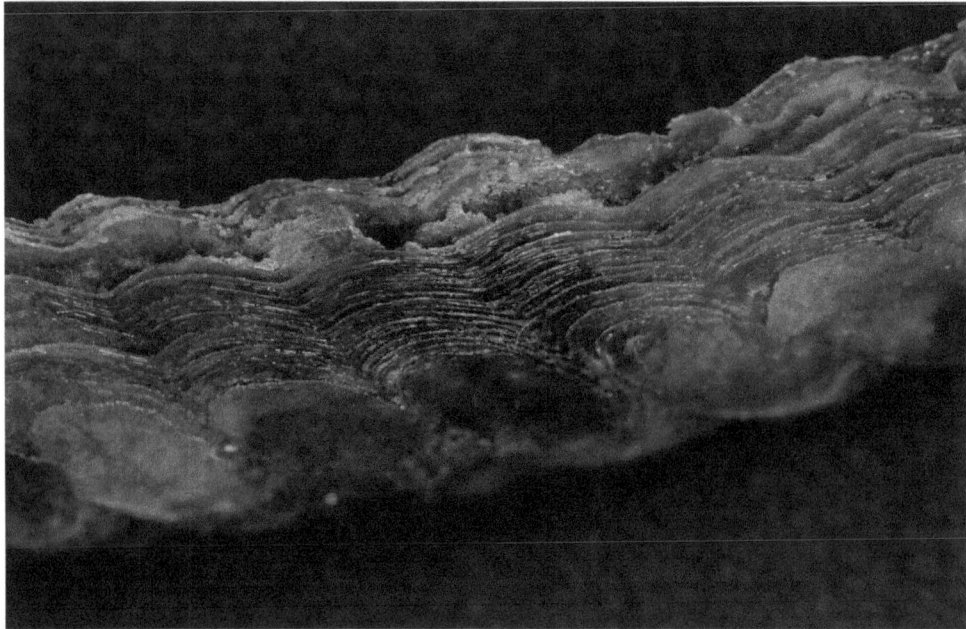

Delaminating layers of keratine formed during the life of the tortoise.

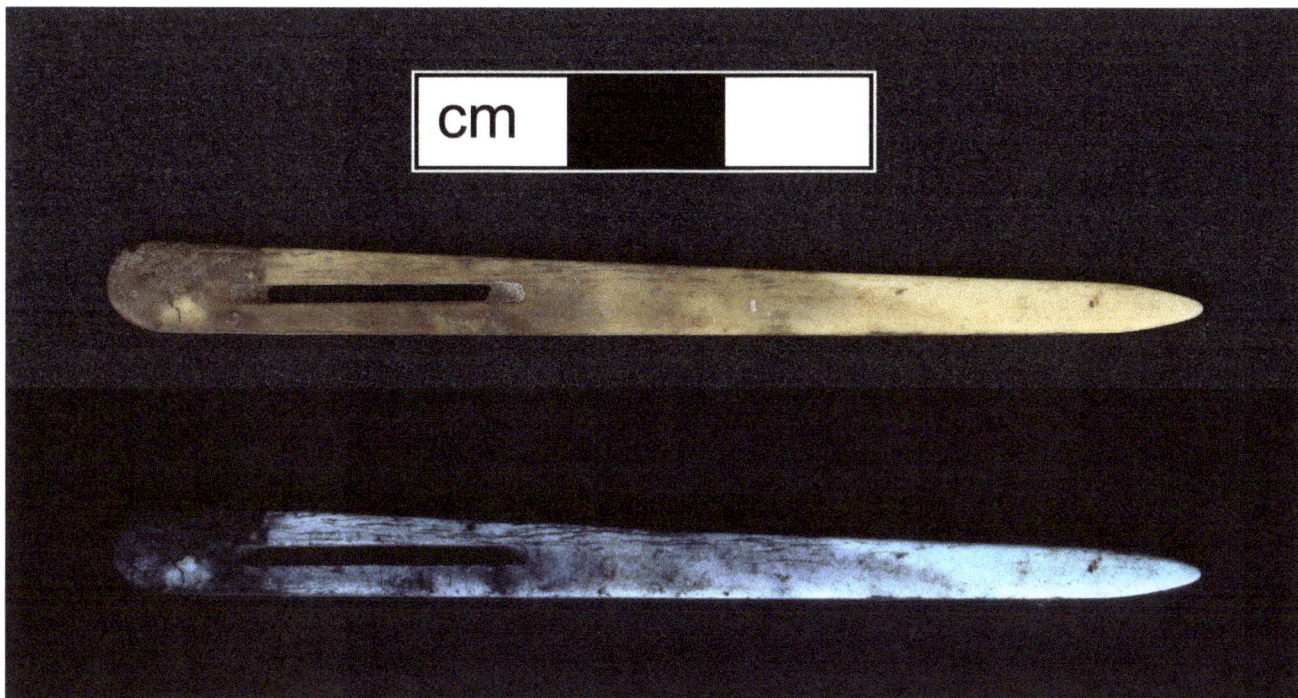

Horn will brightly fluoresce under UV light.

HORN

Horn comes in a range of hues from white to orange to brown/black. Because of the nature of its growth, it is often used to make thin artifacts, such as buttons and combs. New horn is translucent, but this property diminishes with age. Look for surface irregularities, particularly on the artifacts' edges.

BALEEN

Baleen is the comb-like structure that hangs in plates from the upper jaws of certain whales. Due to its plastic properties, it is a versatile and strong material that had a variety of uses until it was replaced by steel and plastics. The color can range from black, gray, and brown, to pale green and cream. It can be difficult to identify visually and can be confused with horn. One key indicator is the tubular layers visible under magnification between the two hard outer plates, though these too can be difficult to identify in degraded archaeological materials.

A fragment of a baleen corset stay.

DETERIORATION OF HORN, TORTOISE SHELL, AND BALEEN

Tortoise shell, horn, and baleen are all made of the protein keratin, which provides these materials with similar plastic qualities when heat is applied. All will fluoresce under UV light with varying brightness and color. All are susceptible to biological attack from insects, bacteria, and fungus. The layered structure of keratin makes it susceptible to delamination.

➤ A hot pin test of keratin produces a burnt hair smell.

➤ To differentiate between the species of origin, mass spectrometry provides the most reliable means of identification as the beta-keratin sequences for cow, sheep, goat, tortoise, and whales are widely recorded.

➤ Observed under UV light, tortoise shell has a chalky blue-white fluorescence on the light portion of the shell, horn is highly fluorescent, and baleen fluoresces different colors depending on the color of the baleen.

Baleen consists of tubules sandwiched between two hard outer plates.

SHELL

Shell used for decorative purposes in composite artifacts is typically white or off-white, and the inner layer of the shell has a characteristic "mother of pearl" iridescence. Shells are made up of layers of calcium carbonate, which are visible under magnification. Highly deteriorated shell, exposed to an acidic burial environment, has a chalky surface that makes identification difficult.

➤ **A chemical characterization test can be used to identify the presence of calcium carbonate using sulfuric acid and observing any telltale effervescence. This will require a small sample for destructive analysis.**

Deteriorating shell, such as this inlaid utensil handle, will exhibit pitting and delaminating of the calcareous layers often seen as watery lines.

Synthetics

EARLY PLASTICS

Casein is a thermoplastic resin produced with milk proteins and formaldehyde; it was developed in the early 20th century. Its ability to take color made it popular for the manufacture of imitation tortoise shell, jade, and ivory for such uses as jewelry, fountain pens, and knitting needles. Nondegraded casein has a brilliant high gloss. When observed under magnification, casein does not have the spotted micro-pattern that makes up the unique tortoise coloration.

➤ **A hot pin test produces a smell of burned milk and the surface will be browned upon heating.**

Celluloid is a nitrocellulose thermoplastic that was a popular imitation ivory in the late 19th century. Oxidation causes yellowing of the surface, not dissimilar from aged ivory. It can be distinguished from ivory by the lack of grains in the surface.

➤ **Caution! Celluloid is flammable. A warm needle test will produce smoke and a camphor/mothball smell. Rubbing the surface with a cloth until warm may also produce the camphor odor.**

Safety celluloid (cellulose acetate) was produced in response to the flammability of celluloid. It is slick to the touch.

➤ **A hot pin test will produce a vinegar odor.**

Catalin is a thermosetting phenol formaldehyde resin developed in the early 20th century as a successor to Bakelite. It is clear rather than opaque and can be brightly colored. It is heavy, hard, and has a greasy feel.

➤ **The hot pin test will produce a formaldehyde odor.**

Bakelite is a thermosetting phenol formaldehyde resin. It is opaque (black, brown, and maroon), heavier than celluloid, feels waxy or slippery, and will not have mold seams. When two pieces of Bakelite and/or Catalin are tapped together, they make a distinctive "clunk" sound.

➤ **The hot pin test produces a formaldehyde odor.**

Lucite is a clear synthetic polymer that gained popularity in the 20th century. It has a slick feel and is lighter than Catalin.

➤ **The hot pin test will produce no smell.**

GLASS

Glass can be characterized by its transparency or translucence when observed with a bright light source. The shape of the object or fragment will often identify the material type, but small fragments such as pieces embedded in jewelry can be difficult to identify. Degraded glass may be distinguished by iridescence or delamination of the surface. Under magnification, glass may contain small gas bubbles from manufacture. Rock crystal, a type of clear quartz, may have small inclusions or flaws. Chips in glass will have a sharp edge and smooth cleave, whereas stone chips are more irregular.

GLASS DETERIORATION

Glass is a construction of silica formed into a loose, semi-crystalline arrangement. The composition of the glass, the ratio of silica to alkali soda ash and lime, is important in the stability of the finished product. Weathering is a form of glass corrosion that commonly occurs in archaeological contexts and results in iridescence or an opaque crust.

The primary variable for the decay of glass is water. Glass is hydrophilic, meaning it attracts and holds moisture. In wet burial environments, the alkali ions are leached away and replaced by hydrogen ions from the water. This creates an increasingly alkali solution of sodium hydroxide ($NaOH$). This alkali solution begins the process of hydrolysis, which breaks down the silica. The loss of silica into this solution then allows leaching of the next layer of alkali ions, and a cycle of deterioration is created. This process is visible as iridescence on the surface of the glass, as the refraction of light between these deterioration layers is no longer homogeneous. Opaque crusts are an advanced stage of weathering. If the decay is extensive, or the glass is suddenly allowed to dry, leaving voids between the layers, delamination can occur. If the burial environment is sufficiently alkaline, the glass can be completely broken down.

In the past, opaque weathering crusts were removed to expose the original color of the glass, but this process removes original material and surfaces.

Iridescence visible under weathering crust.

Burnt glass can undergo devitrification, whereby the semi-crystalline structure reorganizes, and the glass loses its smooth and transparent quality. Some glass will form a whitish crust; others will take on a blue hue caused by the refracted light.

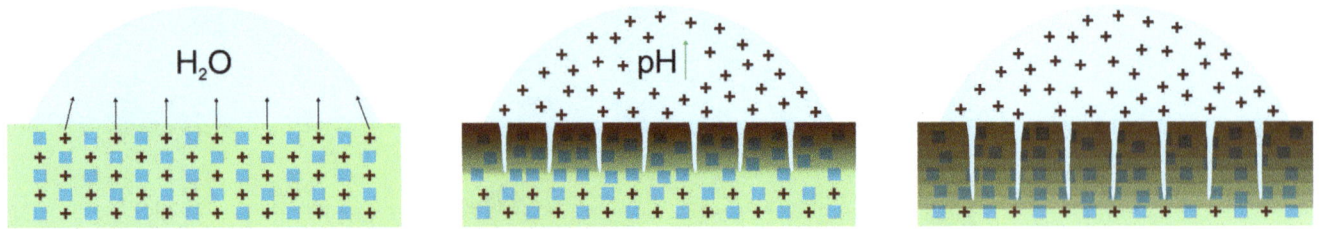

Conservation Terminology for Glass

Consolidation

Consolidation is the use of a dilute adhesive to provide strength to weakened and degraded materials. In the case of glass, it can prevent the further loss of the weathered surface. The need for consolidation should be considered carefully. Consolidation is never fully reversible. Wet or damp glass that may have absorbed soluble salts from the excavation environment must be desalinated prior to consolidation.

Ions leaching out of the glass substrate create an increasingly alkaline solution that further deteriorates and penetrates the structure of the glass, creating a surface composition that is no longer representative of its core material.

CERAMICS

Earthenware, stoneware, porcelain, and brick are all comprised of minerals that have been shaped, dried, and heated to produce a hard, heat-resistant form. Ceramics are the earliest synthetics and one of the most ubiquitous and varied archaeological materials. The many variations and classifications are outside the scope of this guide.

CERAMIC DETERIORATION

Ceramics can include many different mineral components, and the chemical bonds will vary depending on these compositions. Historical ceramics are made using clays, minerals with a flake-like shape that are stacked together, with water acting as a lubricant that allows the clay to be molded. High temperatures drive out that water and create strong bonds between the mineral flakes. Unlike metals, the atoms that make up these minerals are not neatly aligned and are instead amorphous in structure. This is why ceramics break rather than bend when struck.

Ceramic deterioration comes about primarily by mechanical forces. This includes salt intrusion. Low-fired ceramics (earthenware and terracotta, 900–1200 degrees C) are no longer water-soluble after firing, but unlike high-fired ceramics (porcelain, 1200–1300 degrees C) they remain porous to water. In certain burial environments, soluble salts can impregnate the body of the ceramic. As the moisture content decreases after excavation, the salts trapped in the ceramic form crystals that expand and can cause cracking, loss of surface, and disruption of glazes.

Left: Coarse (bottom) and tin glaze (top) earthenware.

Center: Nottingham-type (bottom) and Rhenish blue and gray (top) stoneware.

Right: Chinese winter green (bottom) and Canton (top) porcelain.

Conservation Terminology for Ceramics

Soluble and Insoluble Salts

Salts precipitate out of the groundwater and seawater and are trapped in the porous structure of ceramics, stone, bone, iron, etc. SOLUBLE SALTS (chlorides, nitrates, and sulfates that dissolve in the moisture in the air) are a concern when considering the preservation of artifacts. As the salt crystallizes below the surface of the artifact, the crystals exert pressure in the restricted space and can push off the surface layers of the object. INSOLUBLE SALTS (carbonates, sulfides, and phosphates that require long periods of time to dissolve in water) can form disfiguring crusts that may obscure the surface, but they do not cause physical deterioration and need only be removed for identification or aesthetic purposes.

Soluble salt efflorescence detaching glaze.

Insoluble salts pose less of a danger to ceramics, but can form unsightly crusts that obscure the surface of artifacts.

Field Recovery Methods

Conservation does not begin in the lab but in the field. Often, the earliest preservation decisions are made by archaeologists rather than conservators. The immediate packaging of artifacts in the field can have long-term implications for their future preservation and interpretation. Due to the nature of compliance archaeology and the time constraints and budgets placed on projects, artifacts often remain in their initial field packaging for six months to two years or more. This does not include the time between lab processing and the potential turnover for conservation intervention, which can take much longer.

Deterioration can accelerate as soon as artifacts are exposed, as they try to achieve equilibrium with their new environment. The time during and immediately post-excavation is critical for artifact preservation. It is essential to provide guidance to field and lab crews so that they can make informed decisions about packaging and transporting artifacts from the excavation site, as well as the appropriate washing and storage needs during lab processing. This chapter will recommend specific packaging techniques and lab cleaning methods based on the material types most often recovered in North American archaeological contexts and environmental conditions, with a focus on terrestrial excavations.

Preparation and Supplies

It is always a challenge to predict what resources will be needed for any given archaeological project. What follows is a selection of recommended materials grouped according to their common uses. This is not a comprehensive list, and archaeologists will need to be resourceful as conditions on site develop. Prior to the excavation, the team should stock packaging and lifting materials that best reflect the conditions of a given site and the anticipated material to be recovered. Consult with a conservator to recommend resources and suppliers.

In addition to field supplies, preparing the storage conditions at the processing lab is essential to minimize damage to artifacts during the project. Appropriate conditions can extend the safe storage of artifacts during long processing and reporting periods. Before starting the excavation, provisions should be made to store and care for the different material types expected to be recovered. Make an effort to use packaging materials that will not adversely affect your artifacts if they cannot be unpacked immediately upon arrival in the lab. If artifacts of a certain material type cannot be safely stored, attempt to delay its removal until such arrangements can be made.

A Note on Bags

There are institutional preferences when it comes to the types of bags that are used for material recovery on excavation sites. The two main types of bags used are paper and polythene. Both have positive and negative aspects. Paper bags are affordable, environmentally friendly, recyclable, prevent the buildup of moisture on artifacts, provide some protection from UV damage, and are easy to label with a variety of writing tools. However, they can tear easily, especially when wet or when exceeding their weight capacity; the contents are not easily observed; they may exacerbate mold growth in certain storage environments; they can degrade and become brittle over time, causing disassociation of their contents; and they must be replaced with polythene bags after lab processing. Polythene bags can store wet artifacts or allow artifacts to dry if they are appropriately ventilated; make it easy to view and identify their contents; and can hold heavier material with less chance of tearing.

However, polythene bags are more costly, have a negative environmental impact, and can only be labeled with permanent marker which can rub off. Depending on financial considerations and lab processing procedures, it may be appropriate to consider using a combination of bags. For example, paper bags lend themselves to the transport of most recovered skeletal remains by allowing the material to dry effectively, provide UV protection, and are better able to support the weight of skeletal remains. Polythene bags perform better for collecting larger, heavier quantities of bulk artifacts such as brick and shell. They do not deteriorate in wet environments and provide watertight storage for organic material.

General Packaging				
Object Storage	**Object Transport**	**Labelling**	**Fasteners**	**Padding and Buffers**
Polythene bags	Boxes or plastic totes	Waterproof poly paper (Tyvek)	Stainless steel staples	Bubble wrap
Paper bags	Plastic buckets with lids	Masking tape		Soft and rigid foam sheeting
Tupperware containers		Permanent marker	Plastic coated garden ties	Acid free tissue paper

Wet/Waterlogged Artifacts					
Object Storage	**Object Transport**	**Watering**	**Labelling**	**Fasteners**	**Padding and Buffers**
Polythene bags	Kiddie pools	Garden sprinklers	Waterproof paper (Tyvek)	Cable zip ties	Old blankets, quilts, towels
Plastic mesh or fiberglass screen	Plastic bins	Hoses and pump	Dymo embossing labeler and tape	Stainless steel staples	Soft and rigid foam sheeting
Tupperware containers		Spray bottle or pump spray bottle	Rubber based adhesive tape (Gorilla Tape)		
Plastic sheeting	Plastic bins		Permanent marker	Copper tacks	
Anti-bacterial spray (Lysol)					

Lifting/Consolidation			
Consolidants	**Applicators**	**Support Material**	**Miscellaneous**
Paraloid B-72	Paint brushes	Plywood	Bamboo skewers (soil probes)
PVA	Syringes	Gauze bandages	Plastic cups
	Metal basters	Plaster bandages	Aluminum foil
		Plaster of Paris	Paper towels
		Expanding foam	Tongue depressors (applicator, wedge, probe, handy tool)
		Plastic (Saran) wrap	

Packing by Material Type

IRON

Archaeological iron is one of the most problematic materials for long-term stabilization. When recovered from a terrestrial environment, iron must be kept as dry as possible. For transport from the field, place artifacts in paper or polythene bags. If using polythene bags, be sure that no condensation forms when iron is placed into a bag. Paper bags should only be used for short-term transportation/storage. Small, fragile objects should be placed in individual bags for added protection from mechanical breakage.

LAB PROCESSING: Do not wash iron objects or remove corrosion products. The layers of corrosion are providing passivated protection to the surface layer and any remaining metal. If necessary, loose excavation soil can be removed with a dry natural bristle brush. Do not use wire brushes.

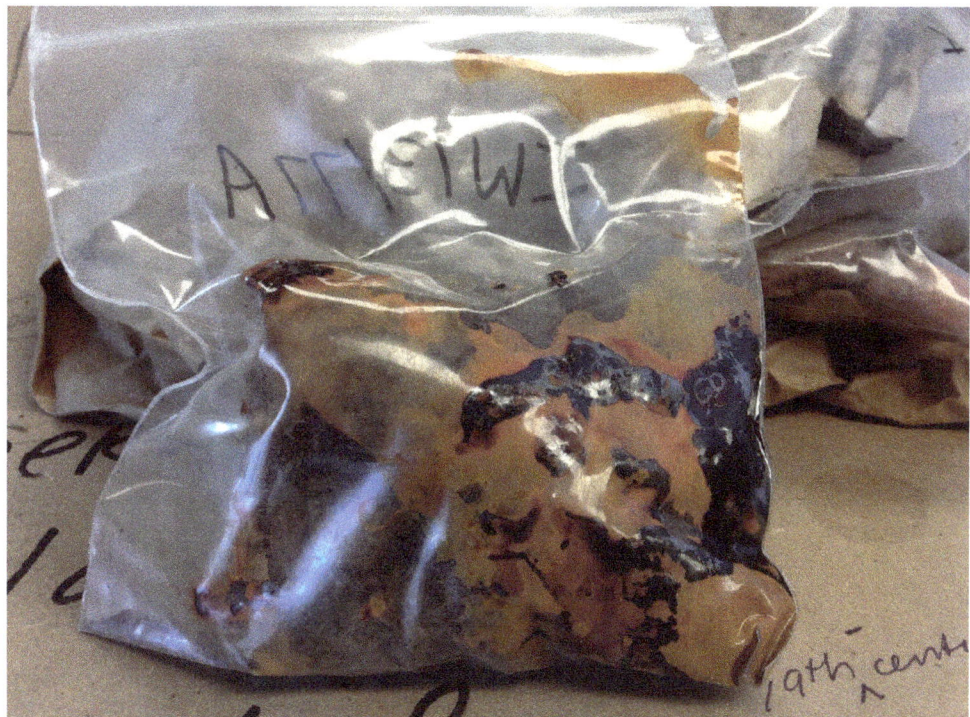

A collection bag that contained wet iron objects which experienced an active corrosion event. The mixed contents concreted into a mass and adhered to the polythene bag.

COPPER ALLOY

Any combination of copper, tin, zinc, and/or nickel (brass and bronze) are considered copper alloys as they share similar material properties. In the field, copper alloys should be allowed to completely dry before they are placed in polythene bags; alternatively, keep the bag open to prevent condensation. Paper bags should only be used for short-term transportation and storage. If there is concern about active corrosion (bronze disease), place a buffer material such as acid-free tissue in the bag alongside the object. This will provide some protection and additional physical cushioning for fragile artifacts. Any observed staining on the acid-free tissue will also serve as an indicator of ongoing deterioration. Do not wrap the object in tissue as the artifacts can no longer be directly observed and may be damaged when unwrapped.

LAB PROCESSING: Do not mechanically remove corrosion from copper artifacts. The layers of corrosion are providing passivated protection to the surface layer and any remaining metal. If the underlying material is exposed without additional conservation measures, it may start a new corrosion event resulting in additional material loss. If necessary, loose excavation soil can be removed with a dry natural bristle brush. Do not use wire brushes.

Acid-free paper can be used to buffer the microenvironment within the polythene bag as well as serve as an indicator of a corrosion event when exhibiting a bluish green stain.

GOLD AND SILVER

Gold and silver objects should be completely dry before they are placed in polythene bags, or else the bag should be kept open to prevent condensation. Paper bags should only be used for short-term transportation and storage. Fragile objects can be supported with padding such as crumpled tissue but should not be wrapped in such a way that the object is no longer visible, as this can cause accidental mishandling and breakage.

LAB PROCESSING: No cleaning is necessary unless it is required for documentation purposes. The soft surfaces can be easily scratched with brushes or picks. If necessary, soil can be removed with a swab moistened with water or ethanol.

WHITE METALS

A white metal alloy includes any combination of lead, tin, zinc, aluminum, and/or antimony. As with other metals, a dry storage environment must be maintained. Because these metals are susceptible to corrosion in low and high pH conditions, the addition of a buffer material (such as acid-free tissue) to a polythene bag may provide some corrosion protection as well as a physical cushion against breakage.

Be aware of handling lead objects, particularly when the surface corrosion is white and powdery. This poses a risk both to the handler and the artifact. Lead is a toxic substance. Minimize skin contact, wear work gloves, and label the contents clearly on the bag to minimize inadvertent handling by lab staff.

LAB PROCESSING: White metal alloys are covered with a thin protective layer of oxide, which provides corrosion resistance. It is important not to disturb that layer when handling or cleaning. Any cleaning, including light brushing, may remove elements of surface detail or decoration that can be revealed with conservation techniques.

CERAMICS

Ceramic artifacts are generally the most robust materials recovered, which accounts for their large recovery volume compared to other material types. There are three notable exceptions to this generalization. Low-fired ceramics can be extremely friable, certain ceramic glazes can be fugitive, and objects from deposits with high salt content may require desalination and careful drying to avoid deterioration post-excavation.

Low-fired ceramics suffer the highest material loss when exposed to water, often in post-processing washing, and from abrasion caused by over-filled storage bags and aggressive brushing. These artifacts should be allowed to dry thoroughly and packed to avoid excessive friction and crushing. If fragments of low-fired ceramic are particularly fragile and require support to be lifted from their excavation environment, consult with a conservator, who may suggest consolidation.

Some glazed ceramics suffer loss of the decorative surface layer. This is often caused by mismatched glaze and body materials, damp burial conditions, or soluble salts from the excavation environment. Except where ceramics are recovered from a wet environment, they should be allowed to dry fully. If salt formation is observed as the ceramics are drying, desalination will be needed. In wet storage, consider spraying the innermost bag with an anti-bacterial product (such as Lysol) to prevent biological growth that may cause staining.

LAB PROCESSING: Wet cleaning is suitable for the majority of ceramic types. Avoid washing low-fired ceramics or glazes that are at risk of detaching from the ceramic body; light, dry brushing may be used for removing excavation soil. Consider consolidation of friable ceramics and fugitive glazes in consultation with a conservator. Artifacts exhibiting salt efflorescence require desalination and controlled drying in consultation with a conservator.

GLASS

Glass recovered from many archaeological contexts in North America has been manufactured using relatively stable formulas. Indicators of unstable glass include delamination, iridescence, spalling, and flaking. Once removed from the ground, all glass should be stored dry regardless of the excavation conditions.

LAB PROCESSING: Bulk glass (such as window glass) can be washed in water to expedite the processing. Do *not* wash diagnostic glass artifacts in water without first testing the stability of the glass. Wash a small fragment and let dry completely. If the glass remains unchanged, proceed washing with water. If the glass turns opaque or cloudy, dry brush or wash with ethanol. If the glass is still wet from the excavation environment, it may be washed in clear, potable water regardless of type.

Wash a small section of the glass with water (left). If the material turns opaque (right), consider alternate cleaning procedures such as washing with ethanol or dry brushing.

LEATHER

Leather is the most common organic material recovered on terrestrial excavations. To lower the potential for microbiological growth, dry leather artifacts should be transported dry and measures taken to ensure that no condensation forms in the bags.

Wet leather has a high conservation potential so long as it remains wet. Use a double-bagging method and fill the inner bag containing the artifact with water. Alternately, spray the interior of the innermost bag with an anti-bacterial spray (e.g., Lysol) and place a wet scrap of cotton cloth or foam in with the artifact. The spray will help to delay microbiological growth. For bulk collections, objects can be placed in mesh bags or perforated polythene bags and then placed in a tub of water. Tyvek paper tags or embossed tags are appropriate for short- to mid-term storage.

Bagged objects should be moved to a cool, dark storage environment to retard mold growth. A refrigerator is the preferred storage environment. Do not freeze wet collections as the water in the object will expand and damage the physical structure of the leather, further deteriorating the object.

LAB PROCESSING: If necessary, clean excavation soil from dry leather with dry brushing. If soiling is particularly stubborn, brush with ethanol (or denatured alcohol which can be purchased in hardware stores) but do not submerge. Wet leather can be cleaned with gentle brushing in potable water. If no conservation is planned, the wet leather may be laid out to dry for long-term storage. Be aware that degraded leather objects may distort as they dry. Drying leather negates the potential for most conservation treatments.

WOOD

Wood can be recovered in many states of degradation. Where a wood object or fragment is in a robust, dry condition, it may be transported and stored in the same manner. Dry wood that is too fragile to be lifted may require additional material support. Consult with a conservator regarding the use of a consolidant. Use of a chemical additive is not recommended where analytical testing may occur, as the use of modern materials will obscure the data.

Wet and waterlogged wood requires wet or damp storage as it will quickly and irrevocably lose dimensional information through shrinkage, warping, and cracking. Small objects can be transported and stored in two nested polythene bags. The inner bag containing the artifact can either be filled with water or the inner bag can first be sprayed with an anti-bacterial spray (e.g., Lysol) and a wet scrap of cloth or foam inserted with the artifact. The spray will help to delay biological growth. Bagged objects should be moved to a cool, dark storage environment, such as a refrigerator, as soon as possible to slow mold growth.

Larger wet objects can be wetted and then wrapped in polythene sheeting. Black sheeting will also provide some UV protection. Individual objects should be double tagged to ensure context information is not lost. Tyvek paper tags or embossed Dymo tags attached with stainless steel staples are appropriate for short- to mid-term storage. For collections that require long-term wet storage or multiple relocations, consider plastic tags (such as livestock ear tags) or embossed Dymo tags attached with brass nails.

LAB PROCESSING: For best results, sampling of wood artifacts for analytical purposes should be carried out as soon as possible. Most interventive conservation treatments, including consolidation and freeze drying, will prevent future analysis. Dry brush only to clean dry wood objects. Wet washing is appropriate for wet and waterlogged wood objects.

Double bag wet organic artifacts with a minimal amount of water and a waterproof paper tag. Biocide can be added to reduce microbiological growth.

SKELETAL MATERIALS

Bone, shell, horn, antler, ivory, and teeth have a wide range of preservation conditions depending on the pH and moisture content of the burial environment. Robust skeletal materials should be transported and stored dry. During the excavation, keep these artifacts out of direct sunlight as the UV can further bleach and degrade any remaining collagen. Ensure that condensation does not form in the bags, as biological growth can easily stain skeletal material. Anti-bacterial sprays are not recommended as they may stain artifacts and interfere with analysis.

Significantly degraded skeletal material may be friable, if dry, or spongy, if wet. Consolidation may be necessary to safely lift and transport damaged objects to the laboratory. Consult a conservator for the best materials to use in these situations. Be aware that the addition of any chemicals for the consolidation of skeletal materials may interfere with certain analytical processes. (Further information on consolidation is in the section on Lifting Techniques.) Consolidation is not appropriate in the excavation of human remains.

LAB PROCESSING: If an artifact was consolidated in the field, it cannot be washed in water. Instead, excavation soil can be removed with wet brushing using water or another appropriate solvent, such as ethanol.

TEXTILE

It is rare to find textile remains relative to other material types. This makes it especially important to care for these artifacts until a research and long-term care plan is developed. Dry textile remains should be kept in cool, dry storage conditions. However, overly dry conditions can cause the fibers to become brittle.

Wet textiles have a high conservation potential so long as they remain wet. Use a double-bagging technique, placing the artifact in the innermost bag in water. Do not place modern cloth in contact with the textile remains as the fibers may snag and cause damage. If necessary, foam can be used for support. If mold growth is a concern, spray an anti-bacterial spray (such as Lysol) on the inside of the outermost bag. Do not let the spray come in contact with the artifact as it can impact the fibers and any dyes.

Textiles are frequently preserved in contact with metal components, such as buckles and buttons. Copper and silver have anti-microbial properties that stave off decay in the immediate vicinity; iron ions migrate into the structure of the textile, replacing the fragile organic fibers with a corrosion casing. Because the textile is the more sensitive material, treat these composite objects according to the need of the textile.

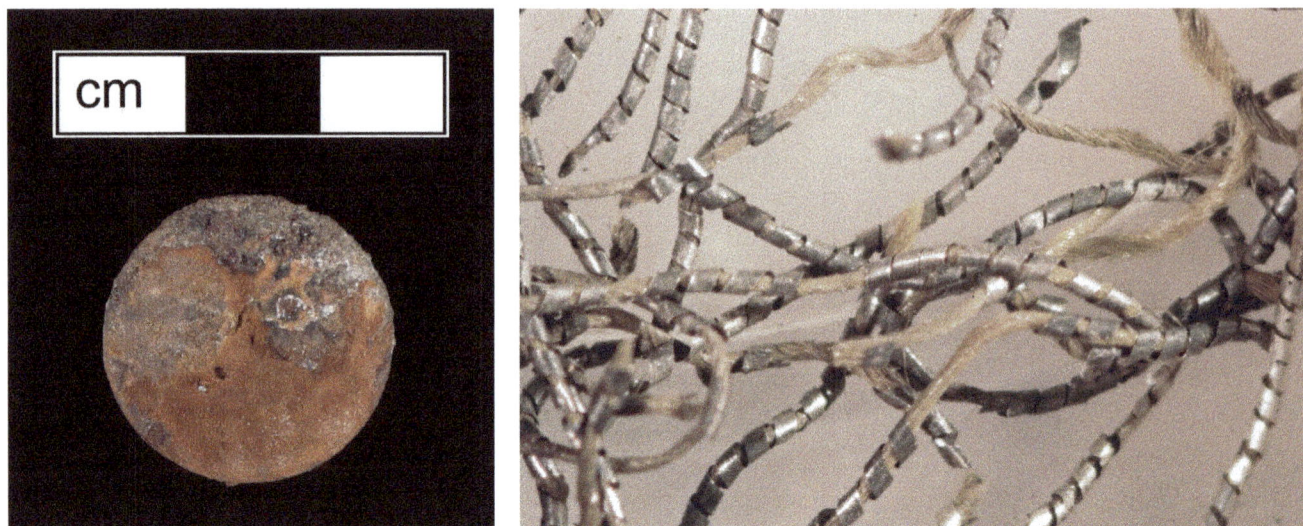

LAB PROCESSING: Dry textile objects may be lightly brushed to remove soil. For stubborn soiling, wet brushing with ethanol is possible so long as no pigments are present and no analysis is expected. Wet textiles should be washed by an experienced conservator. If no conservation is planned, the wet textiles may be laid out to dry for long-term storage. This action will negate the potential for most future conservation treatments.

Left: The fabric covering on this iron button is preserved in mineralized form with iron corrosion products.

Right: The silk core of this silver thread is preserved due to the antimicrobial properties of the metal.

SYNTHETICS

Nineteenth-century sites can produce several synthetic material types including Bakelite, polymethyl methacrylate (PMMA, or acrylic), nitrocellulose, rubber, and other early plastics. Dry and semi-dry artifacts should remain dry; avoid contact with water in lab processing. Polythene bags can be used for transporting objects from the site but will need to be assessed in the lab for signs of active deterioration and off-gassing. For some wet synthetics, there is a risk of distortion upon drying. Maintain a wet storage environment until further assessment can take place in the lab, but do not submerge artifacts in water. Double-bagging or using Rubbermaid containers will provide adequate short-term storage.

LAB PROCESSING: Assess artifacts for instability such as a vinegar or camphor odor, "weeping" droplets, or a blooming, tacky, or crumbling surface. Unstable objects should be isolated and stored in a well-ventilated location, as a build-up of these chemicals in a micro-environment may accelerate their deterioration or cause damage to neighboring artifacts. Refrigerated storage will slow deterioration. Wet objects may be gently cleaned with wet brushes, but dry artifacts should not be exposed to water and dry brush only. Wet objects may be slow-dried in a controlled setting. If working with a large quantity of wet synthetics of the same type, dry a sample before proceeding.

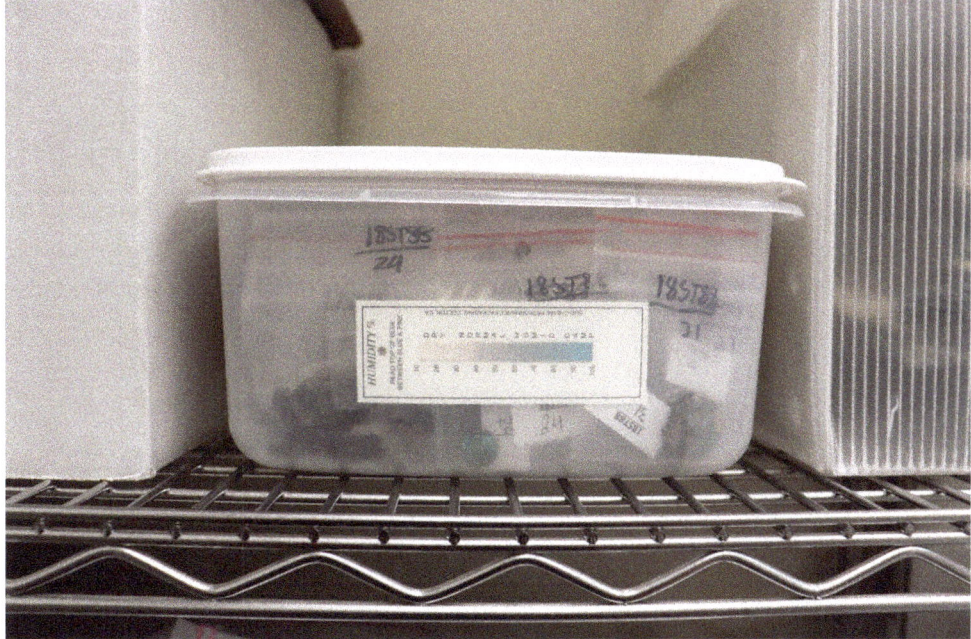

Placing an RH strip in metals storage containers (particularly when used in combination with a desiccant like silica gel) is an easy way to monitor the microenvironment.

COMPOSITE ARTIFACTS

When recovering artifacts that are composed of multiple material types, the general rule is to address the needs of the most reactive material. For example, an iron knife with a bone handle recovered from a dry environment should be packaged and stored according to the needs of the iron. If the iron starts to deteriorate, it is likely to damage the bone in the process. The excavation environment also plays a role in determining the priority of preservation. Preservation of a leather shoe with iron hobnails from a wet environment should focus on the leather. Although iron prefers a dry state for its preservation, uncontrolled drying of the leather is irreversible. However, preservation of the same shoe from a dry environment may need to prioritize the stabilization of the iron, since the leather has less potential for conservation in its dried state.

Wet Artifact Recovery

When preparing wet artifacts for transport to the laboratory for processing, be mindful that wet storage can be both heavy and messy. Find measures to reduce the amount of water conveyed with the artifacts as much as possible. Small artifacts may benefit from double bagging, where the inner polythene bag containing the artifact also holds water. Unless you are using water to provide buoyancy to a delicate object, a small amount of water will suffice. Groups of larger objects or crates of objects can be placed on a tarp and wrapped together after they have been thoroughly wetted. Individual over-sized objects may be wrapped in black plastic sheeting, cut to measure, and secured using cable ties. Label all tarps and sheeting with labels made with a rubberized tape (such as Gorilla Tape) and permanent marker.

Water reservoirs can be added to these storage envelopes by placing wet foam or textiles in with the artifacts. This is especially useful during transport or for short periods when staff are unable to access the materials to monitor the moisture content. This is not a long-term storage solution, as microbiological growth will quickly form in this environment and moisture content cannot be maintained indefinitely.

Once artifacts arrive at the laboratory for processing, mid- and long-term storage environments are required. Small artifacts that can be stored in polythene bags or plastic containers should be placed in refrigerators or other cold storage. Temperatures below 40 degrees slow biological growth. Do not allow wet collections to freeze; the water in the cells and pores will expand and damage the physical structure of the wood, further weakening the condition of the artifact.

Mesh bags can also be used to secure wet organics. These can then be stored in tubs or tanks, which makes it easier to change or filter the water.

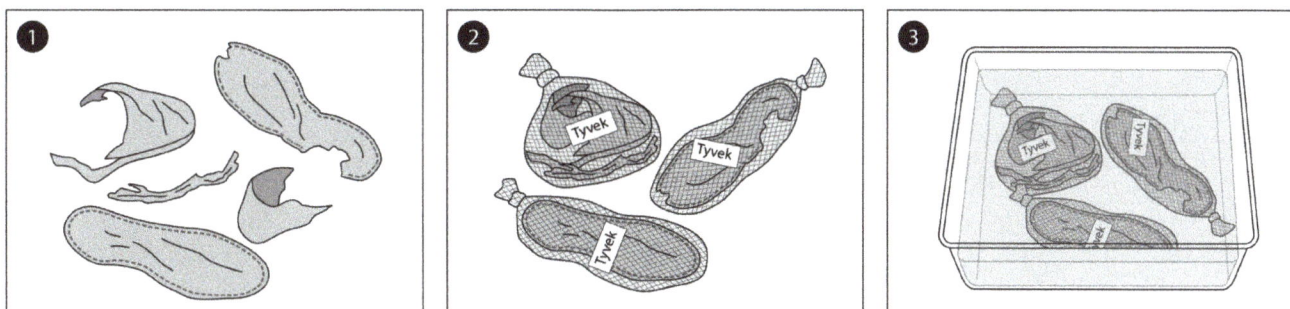

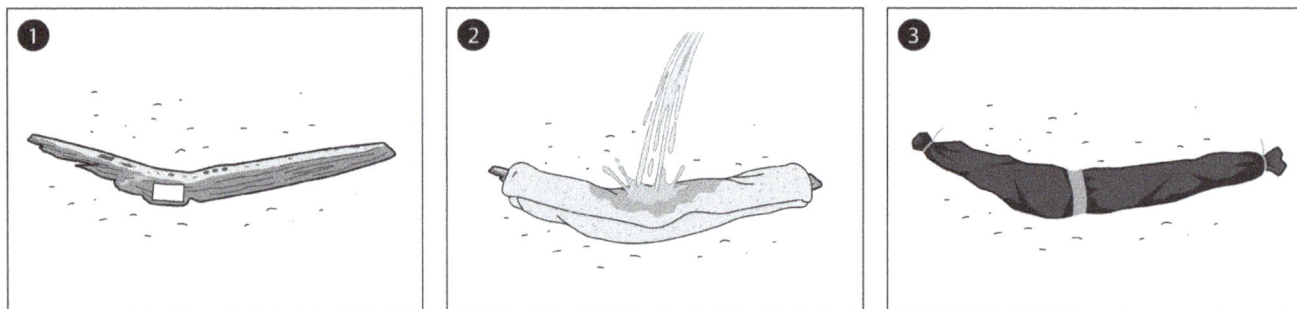

For mid-term storage of large and oversized wet materials, a folded tarp allows for the contents to be wetted on a regular basis. Another mid-term storage environment, which allows for submersion, uses tarps or plastic sheeting supported by sandbag/cinderblock walls to create a shallow pool. Long-term storage requires that the artifacts be completely submerged. Many different storage containers may provide wet storage, including livestock tubs, lined wood crates, and above-ground pools. Avoid the use of metal tanks, unless lined, as they can corrode and deposit minerals on the artifacts that may lead to future preservation issues. Long-term storage tanks will require filters, bubblers, and pest mitigation, depending on the duration of use. Biocides should only be used in consultation with a conservator, as such chemicals may have health and safety complications that can impact future access.

It is not necessary that the water used to store artifacts come from the same water source where the artifacts were recovered. If the water available locally is high in mineral content, consider using distilled or reverse osmosis water. Do not use deionized water as this may leach minerals from your artifacts.

Wrap wet timbers in opaque plastic secured with zip ties at either end. Close any gaps in the plastic with water-resistant tape. If there is a concern for moisture loss, include a wet towel within the package.

Above-ground pool as temporary storage for wet timbers with bubblers agitating the water to prevent mosquito infestation.

When considering long-term wet storage, be aware of the challenges in controlling biological growth, such as insects. Best efforts should be made to keep the contents cool and away from sunlight. Changing the water regularly (more frequently during the warmer months) is a good preventive measure. If this is not practical, circulating and filtering the water is recommended. UV filters can also reduce algae formation. Mosquitos are not harmful to the artifacts but can be frustrating to staff. Keep the storage containers covered if possible. Where this is not practical, consider agitating the surface of the tank with a pump or adding mosquito dunks to prevent larvae growth.

Lifting Techniques

Support lifts are used to remove and transport fragile artifacts from the field. Lifts can also be used for keeping fragmentary vessels intact, so that their contents can be excavated off site in a more controlled environment or x-rayed to determine the contents prior to their excavation. There are five main types of support lifts commonly used in the field: block lifts, wrap supports, pedestal lifts, facing/backing, and consolidation. Consult with a conservator for guidance on which lifting techniques and materials may be best suited for a particular situation, or if possible, have a conservator perform the lift themselves.

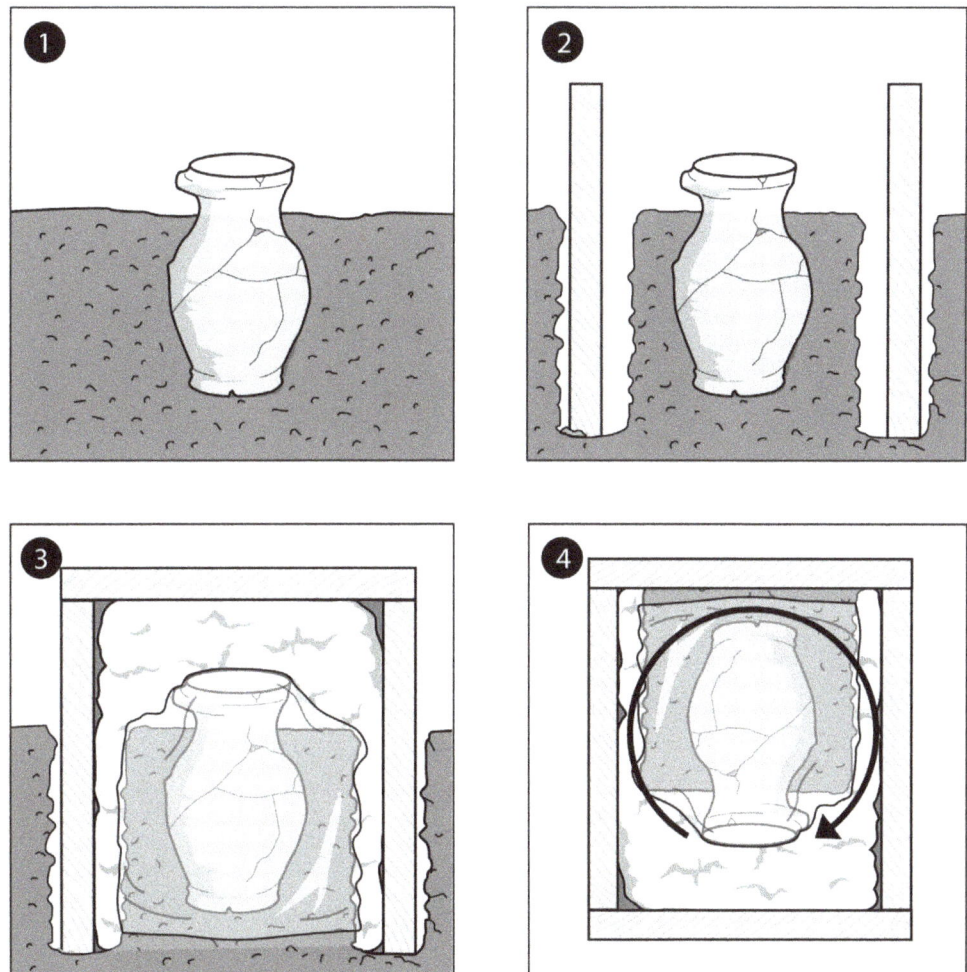

FIGURE 1

1. Identify object. **2.** Estimate the dimensions and excavate to provide space to wrap or box around the object. **3.** Line the interior of the box with a protective, waterproof layer, fill with foam or plaster, and close the box. **4.** Detach the soil block and turn over to crate for transport.

Block lifts (Figure 1) incorporate the use of the surrounding matrix to support the object. They cut into the soil around an artifact and a container/support is constructed around the object and remaining soil. As the surrounding contexts are removed down to the base of the box/artifact, the bottom can be undercut and the entire package removed. Block lifts are useful in protecting the contents of the lift, as the soil acts as both a support and buffer material. However, for larger objects, the quantity of soil can make for very heavy lifts. Keep in mind there are few cases where the condition and dimension of particularly large objects can be correctly determined so that a box may be accurately constructed; also, not all projects may accept cutting into unexcavated contexts.

For larger object lifts, a **wrap support** (Figure 2) may provide a more viable alternative. Wrap supports are also very useful for the removal of fragmentary vessels and their contents. As the object is exposed over the course of the excavation, bandages are wrapped around the object and even and firm pressure is applied; this presses against the soil and the contents within the object to provide support. These bandages can be made of fabric,

FIGURE 2

1. Firmly tie wrapping material around the exposed section of the object; do not remove interior soil. **2 and 3.** Continue to firmly wrap as more of the object is exposed. **4.** A counter wrap may be needed to ensure even pressure. **5.** Support object from beneath, as the base is often the weakest point. **6.** Wrap vertically to the base support material for transport.

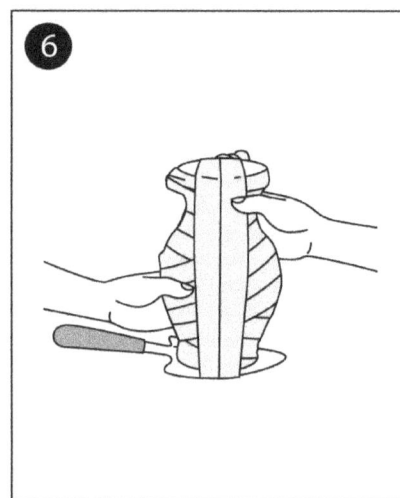

plastic, plaster, or a combination depending on availability and the needs of the object. If plaster bandages are desired to provide additional rigidity and strength, first apply a barrier layer, such as plastic wrap, to prevent the migration of plaster into the object and facilitate later removal. If the object is a wet organic artifact, using wet fabric bandages can help the artifact retain moisture; then apply plastic wrap to slow evaporation. Wrap supports are useful in situations where a fragile artifact cannot be immediately removed. They also allow for the systematic excavation of the surrounding area.

Artifacts that have a low profile can be removed using a **pedestal lift**. (Figure 3) As the object is exposed, continue to excavate beyond its base until it has been raised up by a small pedestal of soil beneath. Surround the object with an open box and line the interior with a barrier layer of plastic, covering the object and as much of the box interior as possible. Some loose excavation soil can be added on top of this layer to hold the plastic film in place against the object. The empty space is then filled with a solid support material such as plaster or expanding foam. When this support material is firm, the soil pedestal can be broken, and the lift can be turned upside down for removal.

A **facing/backing** (Figure 4) support can be used for low profiled artifacts. This is especially useful when an object is very fragmented and there is a need to maintain the orientation of the pieces. Adhere strips of bandages in perpendicular layers over the exposed surface of the object. Be sure to cover the top and sides of the object. Bandages can be made of fabric or of paper with

FIGURE 3

1. Continue to excavate around the object until it is raised on a pedestal of soil. 2. Surround object with a box, line the interior with a waterproof layer, and fill with plaster or foam. 3. Close the box lid and detach the soil block. 4. Turn over for transport.

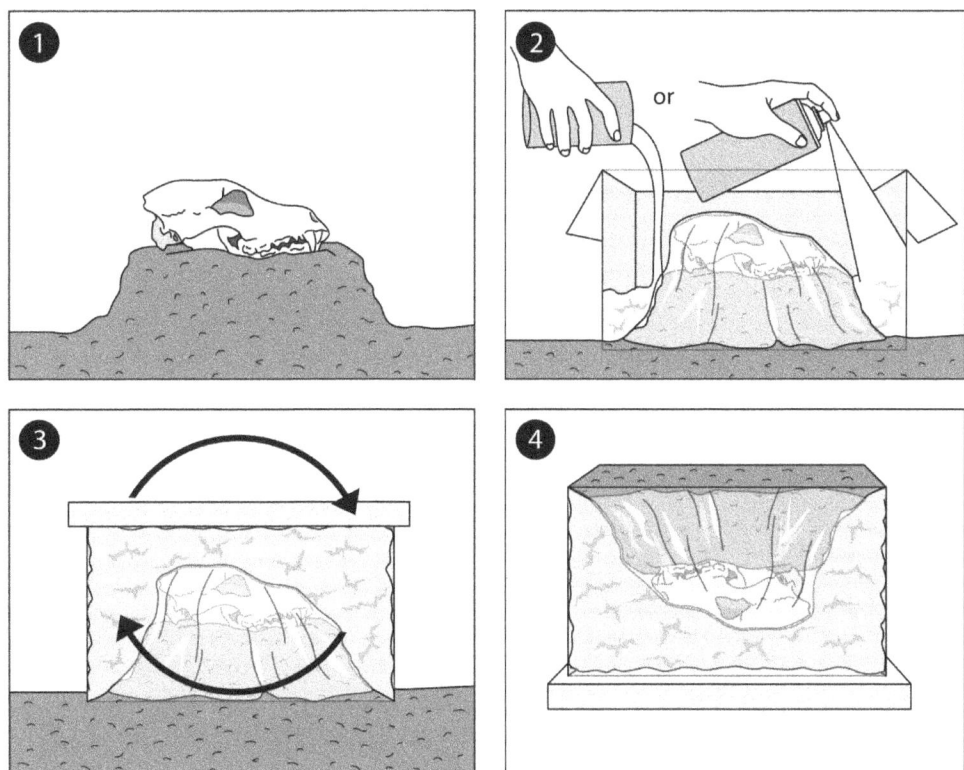

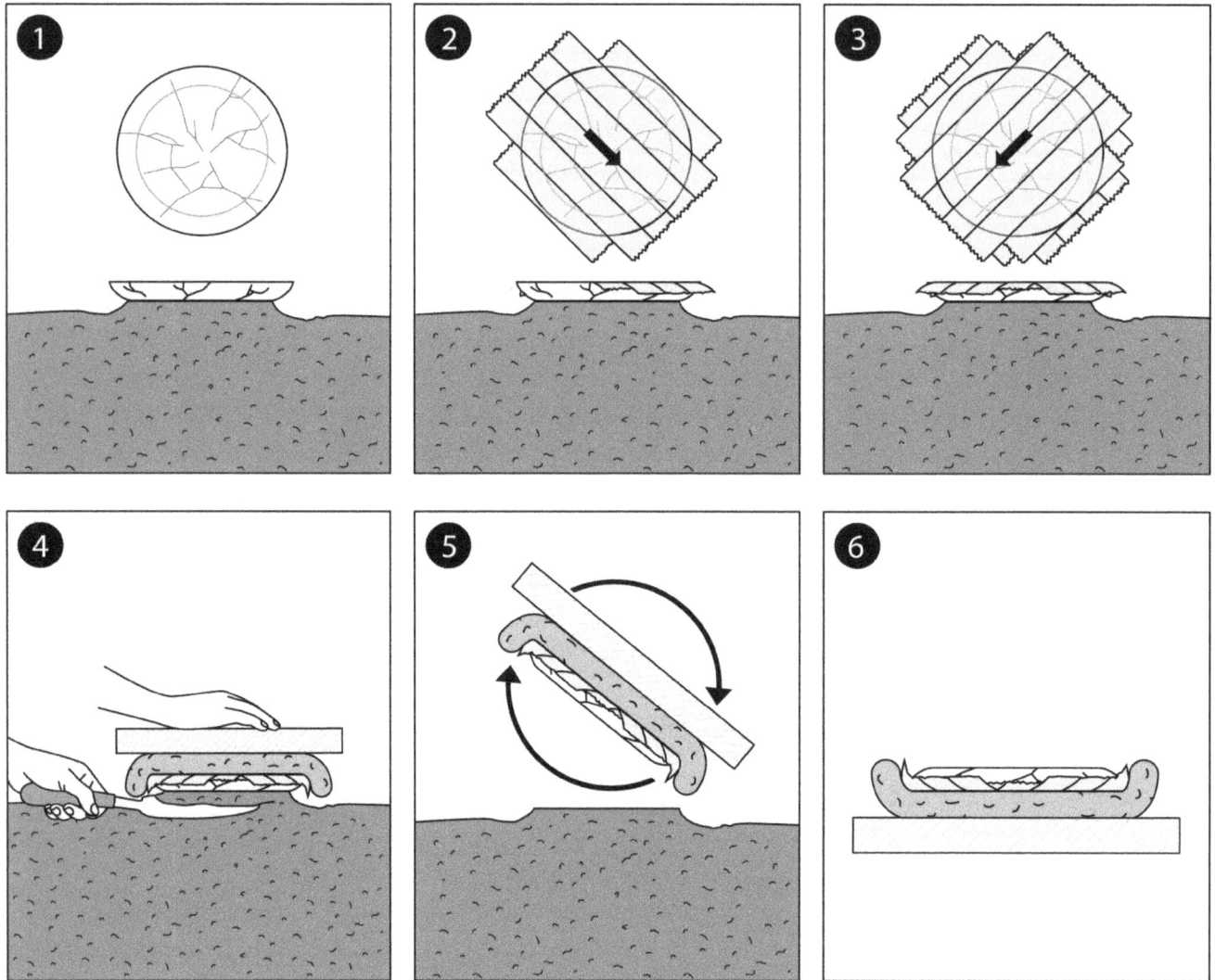

adhesive or plaster strips. Consider a barrier layer between the bandages and object, depending on the material being lifted; think about how the adhesives can be later removed. Once the layers of bandages have completely dried, use a padded layer (this can be a folded towel or a polythene bag filled with soil, etc.) to support the top of the artifact as the soil from beneath the object is loosened. Using a tool to support the underside of the object, flip the package over so that it is being supported on the padding and place on a firm, stable surface.

Collar lifts (Figure 5) are a form of block lift in which the removal of the surrounding soil serves as the mold for a plaster or foam collar that supports the soil and object as it is excavated in place of a prebuilt support. This method can be very useful in damp excavation conditions. Where some lifting materials may not hold up in wet environments, expanding foam requires moisture to be most effective.

The previous lifting methods are examples of structural supports. **Consolidation** (Figure 6) provides support by impregnating the artifact and/or the surrounding soil with a dilute adhesive to improve the physical

FIGURE 4

1. Clear soil around the sides of the object.
2. Place barrier layer if needed and apply adhesive or plaster strips across the entire surface. **3.** Apply strips in alternate orientation.
4. When dry, place padding and ridging support on object and detach the soil block.
5 and 6. Turn over for transport.

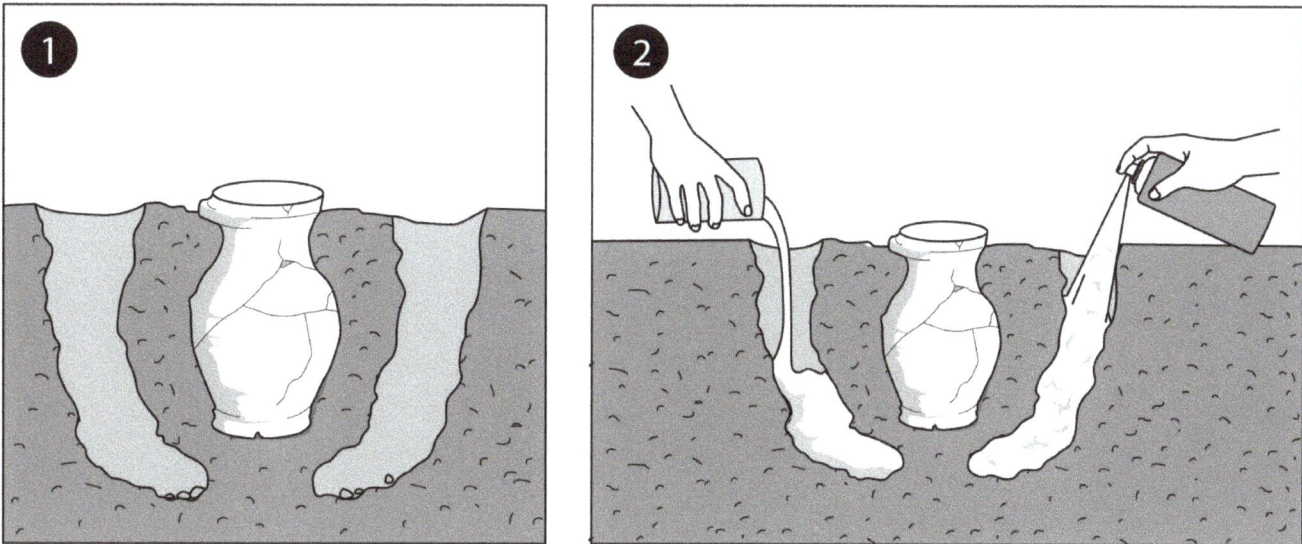

FIGURE 5

1. Excavate around and slightly under the object. **2.** Fill with plaster or expanding foam.

strength of the object. It can be used alone or in conjunction with other lifting methods as needed. The adhesive and consolidant used are determined primarily by the artifact material, the moisture content of the excavation environment, and whether the consolidant will require later removal. Paraloid B-72 and polyvinyl acetate (PVA) are the most common consolidants used on site. A conservator should be consulted to assist in selecting a consolidant. Each has limitations: for example, Paraloid B-72 cannot be used in wet or damp conditions, and PVA should never be applied to metals.

CONSOLIDATION USING PVA

PVA is a medium-strength adhesive that is water soluble and widely available, making it a preferred consolidant for many conservators and archaeologists.

PVA can be mixed with water; however, this will greatly increase the drying time. Instead, a 5–15% solution in ethanol or denatured alcohol is recommended.

Consult with a conservator to determine if PVA consolidation is suitable.

1. Clear the surface of the artifact and the surrounding area. Take a photograph to document the object before consolidation.

   If the artifact is too delicate to fully expose, excavate around the artifact so that it is supported by a pedestal of soil, which will be consolidated with the artifact to provide structural support.

2. Saturate the artifact (and soil) with ethanol, using a spray bottle or pipet. For larger artifacts, a second application may be required to ensure saturation.

3. Apply a coating of PVA to the artifact (and soil) with a spray bottle or pipet. After an hour, apply a second coat. Take photographs to document the process.

4. After one hour or when dry, loosen the soil beneath the artifact and lift, supporting it with a trowel or flat board.

For large artifacts, consolidation may be used in combination with other lifting techniques.

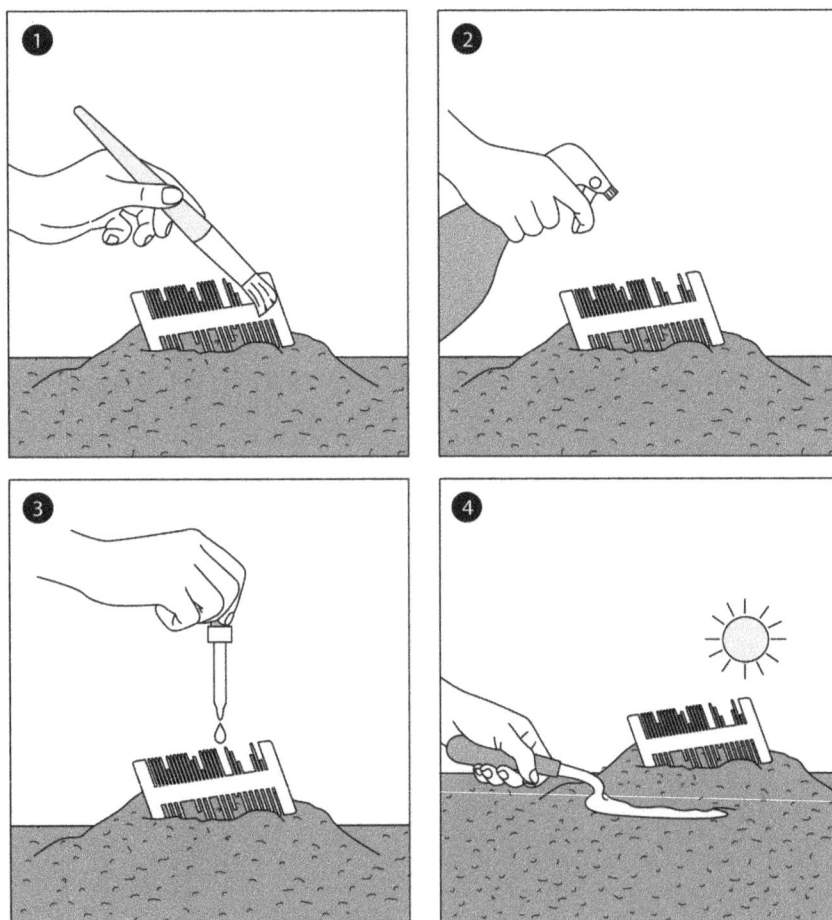

FIGURE 6

Consolidation.

Notes

If laboratory grade PVA has not been included among the excavation supplies, it can be replaced with Elmer's Glue-All in emergency situations. Elmer's Glue-All is an emulsion of PVA, polyvinyl alcohol and propylene glycol.

PVA can be cleaned from tools using ethanol or hot water.

When planning a lift, consider the following:

- Is a lift necessary? Or can the object be recorded in situ and removed as fragments?
- What materials will be coming in direct contact with the artifact? Do they have the potential to cause damage? Can they be "easily" removed?
- How long will the artifact be stored within the lift materials?
- Do the artifacts need to be stored wet or dry? Are there other special storage requirements?
- How might the weight of the lift impact removal and transport from the site? Can a cardboard box replace a heavy wood frame? Could expanding foam be used in place of plaster?
- Could you combine lifting techniques (Figure 7) depending on the fragility of the object? For example, should you apply a facing prior to a pedestal lift?

FIGURE 7

Combination lift example. **1.** Excavate around and slightly under the object. **2.** Firmly tie the wrapping material around both the object and the soil block.

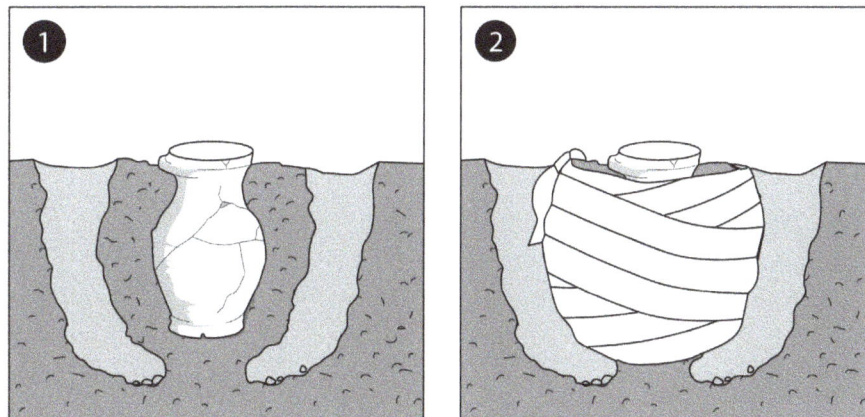

When choosing a support material, remember that:

- Plaster and expanding foam are easy to stock with field supplies or are readily available commercially.
- Plaster is heavy but stable and will not react with the artifact. Plaster dust may migrate into porous objects if you are not careful.
- Expanding foam is light and easy to remove; however, it off-gases chemicals that can negatively impact metal and organic artifacts and should only be used for short-term supports. When purchasing expanding foam, look for dispensers that allow for multiple reuses.
- In the field, a variety of other packing materials may be utilized based on availability. Keep in mind that the chosen material should not compress over time and must be removeable.

| | Organic | | | | | | Synthetic | Inorganic — Metals | | | Inorganic — Non-Metals | |
	Skeletal	Faunal	Leather	Wood	Floral	Textiles	Rubber/Plastic	Iron	Copper Alloys	White Metals	Glass	Ceramics
Dry	Dry Packing; Ensure No Condensation; Keep Out of Direct Sunlight						Dry Packing, Ensure No Condensation; Use Support Material for Fragile Artifacts; Include Buffer Material with Any Signs of Active Deterioration *(Dry)*					
Wet (Terrestrial)	Dry Completely Before Dry Packing; Ensure No Condensation; Keep Out of Direct Sunlight		Wet Packing If Double-Bagging, Spray Interior of Inner Bag with Anti-Bacterial Spray			Wet Packing; If Double-Bagging, Spray Interior of Outer Bag with Anti-Bacterial Spray	Wet Packing; Do Not Submerge	Dry Completely Before Dry Packing; Ensure No Condensation; Include Buffer Material with Any Signs Of Active Deterioration *(Wet [Terrestrial])*				Dry Completely Before Dry Packing
Waterlogged	Wet Packing/Dry Pack Consolidated Artifacts; NO Anti-Bacterial Spray							Wet Storage *(Wet [Marine])*				

	Organic					Inorganic				
	Faunal		Synthetic	Floral		Metals		White Metals	Non-Metals	
	Skeletal	Leather	Rubber/ Plastic	Wood	Textiles	Iron	Copper Alloys	White Metals	Glass	Ceramics
Dry	Wet wash, UNLESS consolidated in the field.	Wet wash, UNLESS consolidated in the field.	Dry brush.	Dry brush. For stubborn soiling, wet brush with ethanol unless pigments are present.	Dry brush. For stubborn soiling, wet brush with ethanol unless pigments are present.	Dry brush to remove soil. Do not use a metal brush. Do not remove corrosion.	Dry brush to remove soil. Do not use a metal brush. Do not remove corrosion.	No not wash or brush.	Test diagnostic glass prior to wet washing or wash with ethanol.	Wet was unless low fired or has a loose glaze
Wet (Terrestrial)	Wet wash, UNLESS consolidated in the field. Store dry.	Wet wash. Store wet, cool, and dark.	Wet wash. Store wet, cool, and dark.	Wet wash. Store wet, cool, and dark.	Wash by conservator. Store wet, cool, and dark.	Wet wash.	Wet wash.	Wet wash.	Wet wash.	Wet wash.
Waterlogged	Consult with conservator regarding consolidation and/or controlled drying.	If not a candidate for conservation or analysis, can be laid out to dry for long-term storage.	If not a candidate for conservation or analysis, can be laid out to dry for long-term storage.	If not a candidate for conservation or analysis, can be laid out to dry for long-term storage.	If not a candidate for conservation or analysis, can be laid out to dry for long-term storage.	If not a candidate for conservation or analysis, can be laid out to dry for long-term storage.	If not a candidate for conservation or analysis, can be laid out to dry for long-term storage.	If not a candidate for conservation or analysis, can be laid out to dry for long-term storage.	If not a candidate for conservation or analysis, can be laid out to dry for long-term storage.	If not a candidate for conservation or analysis, can be laid out to dry for long-term storage.

Color legend: Dry | Wet (Terrestrial) | Wet (Marine)

Conservation and Collaborations

The Whys and Hows of Conservation

HOW TO INTRODUCE CONSERVATION TO THE UNINITIATED

While many developers are reconciled to the requirements for archaeological survey and mitigation, their understanding of the perpetuity of this responsibility may be less solid. It is our responsibility to explain that fieldwork is only one component of the archaeological process. Of equal importance is the lab processing, research, analysis, reporting, conservation, and curation of data and artifacts. Archaeology is a destructive process. The reports generated and the artifacts gathered are often the only remains of a site. While the information retained in the site report may be copied, distributed, and saved in multiple formats, the artifacts that support this data are much more vulnerable to loss and damage. Recovering an artifact from a stable burial environment only to have it deteriorate in storage goes directly against the intentions that drive archaeological mitigation efforts.

When bid proposals do not include conservation and curation budgets, clients can be surprised by these additional costs further into the project. It is much better to set these expectations in early planning phases. This can be more difficult to justify to developers, whose main concern is minimizing expenses, than to federal and state agencies that are more invested in long-term resources. In these cases, it may be necessary to explain that conservation is part of their Section 106 responsibilities and to emphasize the use of exhibitables to boost public perception of the company or project.

CONSERVATION BUDGETING

When budgeting for conservation, it is important to understand the types of artifacts that will be encountered on the site, the amount of material likely to be generated, and the state of preservation. Use Phase I artifacts to assist with estimates for Phase II- or III-level excavations. The type of site will heavily influence the potential need for conservation. Pre-contact sites generally require less conservation intervention due to the absence of metal artifacts. The later the site, the more metal is likely to be recovered.

Seventeenth-century sites tend to be 5-10% metal, 18th-century sites are closer to 10–20% metal, and 19th- to 20th-century sites tend to be higher still, at 20–30% metal. These percentages probably vary by region and certainly vary by site type (i.e., the estimate is higher for an iron furnace, lower for a ceramic kiln). Urban sites will typically have larger assemblages than rural sites. Wet excavation sites and clay soil matrices have a higher potential for organic preservation and potential conservation intervention.

Not all artifacts or collections require post-recovery conservation efforts. This will depend on the site and types of materials recovered. There are many preservation and reconstruction methods available to conservators, but for the purposes of Section 106 compliance, conservation only need focus on the chemical and physical stabilization of collections. On average, only 1–2% of an assemblage may require some sort of conservation, either chemical intervention or environmental controls, for long-term storage. Even a small budget will ensure that the most critical artifacts receive attention. The exception would be on sites where there is a high degree of organic preservation.

If possible, have a conversation with a conservator in advance of the project to assist in the budgeting process. Conservators should be open about their cost structure. If there isn't a pre-established relationship with a conservator, an alternative strategy would be to match the conservation budget with the funds allocated for site/collections analysis (i.e., faunal, soil, lithic, or ceramic).

Material Object Type	Estimated Labor Hours (Dry)	Estimated Labor Hours (Wet)
Small Iron Objects (spikes, hardware, etc.)	2–5	
Medium to Large Iron Objects (fireback, cannonball, kettle, etc.)	6–21	
Large to Extra-Large Iron Objects (cannon, anchor, etc.)	49–185	
Small Non-Ferrous (copper, lead, white metal) Objects (coins, buckles, household items)	2–7	
Glass and Ceramics—Stabilization of Sherds	2–3	
Wooden Objects	3–13	3–23
Small Leather Objects	2–3	2–9
Small Textile Object	2–3	2–9
Bone / Shell / Horn	2–4	2–5
Large Composite Object (typically iron or wood)	12–50	12–50

Time estimates are based on previous experience and should be used for planning purposes only. Specific artifacts may require more or less time, depending on individual differences such as degree of corrosion or concretion, object size, moisture content, etc. Some artifacts are able to be treated in batches (i.e., desalination of ceramic sherds or small iron objects), which can reduce overall labor and allow for more cost-effective treatment of the artifacts.

WHAT ARE CONSERVATION COSTS?

It is often infeasible to have a conservator working in-house. Instead, these tasks are outsourced to conservation facilities or conservators in private practice. Given the nature of the work, conservation is heavily labor driven. The specialized background of conservators and the many hours that are needed to carry out preservation tasks can be expensive.

Most conservation work consists of mechanical processes whereby a conservator is actively removing concretions, testing solutions, applying consolidants and coatings, mending breaks, etc. There are chemical treatments that can be applied to bulk collections to reduce labor hours and cost. However, most chemical treatments must be closely monitored as they can quickly cross over from aiding the treatment process to causing irreversible damage. Mechanical treatments are generally preferred as they are more easily controlled, and the conservator is more likely to observe features that may be of interest or importance to the archaeologist and curator.

Conservators are also diligent in recording the progress of treatments. Just as it is necessary to take detailed notes, measurements, and photographs during an excavation, it is equally important to document conservation treatments, as the process is often irreversible, and damage can occur to particularly vulnerable objects. It is also useful for future conservators and curators to understand past treatments should an artifact need to be utilized for display, further analysis, or re-treatment.

In addition to labor, specialized tools and materials may be needed for certain treatments. And some of the chemicals used are considered hazardous, needing additional considerations for both user safety and disposal of waste. Labs as well as conservators in private practice must also hold specialized insurance that contains property and professional liability coverages.

Certain conservation treatments or material types are more expensive than others. Cost will most often depend on the amount of time needed to carry out the work, the price of the materials, and the use of specialized equipment. For example, treatment of a white metal artifact is generally less expensive than a comparably sized iron artifact. Iron often takes longer to clean due to the nature of the concretions and may require desalination, which uses additional chemicals, monitoring equipment, and more hands-on time. Treatment of waterlogged organic objects is particularly expensive because the chemicals used to preserve the material have a high cost and often require the use of specialized equipment, such as a vacuum freeze dryer.

FINDING A CONSERVATOR

For most excavations, it is impractical to have a conservator in the field. However, when an issue arises, it is important to know who to call. A conservator may be asked to provide training to field crew, give advice over the phone, assist with identifications, recommend specialized analysis, recover fragile or over-sized artifacts, and assist with conservation budgeting. An experienced archaeological conservator can be a resource to connect archaeologists, educators, researchers, and curators. Finding that conservator can be a process.

The American Institute for Conservation (AIC) has an online tool to search its membership. It can sort this roster by location and specialty. AIC only lists those professional members who have completed a peer review process. Note, however, that some conservators may not be able to take on work from outside their institutions.

State Historic Preservation Offices (SHPOs) are a good resource for identifying local conservators. The compliance and review staff will often have experience working with archaeological conservators and can provide recommendations. This is also true of the curatorial staff at archaeological repositories.

CONSERVATOR QUALIFICATIONS

It is highly recommended that conservation staff have a master's degree in conservation from an accredited university program. There are universities that provide classes, workshops, and certificates in conservation methods. These are *not* the equivalent of a master's degree, and their recipients should not be considered professional conservators on those qualifications alone, just as a field school season does not make an archaeologist. Post-graduate programs not only provide instruction on the methods of conservation, they also provide the training to evaluate all potential treatment options, judge the best course of action (or inaction), and recognize when to stop treatments. Conservation and archaeology are destructive processes. An experienced conservator can recognize a preserved surface amidst the iron corrosion products in the same way a professional archaeologist is able to differentiate stratigraphy by soil color and texture. Without the requisite training and experience, it is easy to destroy or misinterpret information.

Professional conservators are also trained to adhere to ethical considerations when evaluating collections for conservation intervention. The AIC

Code of Ethics and Guidelines for Practice describes the standards to which conservators approach treatment and advocacy for the care of cultural property. This also includes continuing professional development to keep informed of the newest preservation standards and technologies.

Archaeological conservation is a subcategory of object conservation. There are many crossovers in treatment applications, as the chemical makeup of historic objects and archaeological artifacts are similar. However, the burial conditions of archaeological materials can present additional complications to their preservation. If a project is looking to hire a bench conservator (someone who is not expected to assist with the field recovery or post-excavation analysis), an objects conservator with little archaeological experience may be perfectly suited to the task. The exception would be if waterlogged organics or actively corroding metals are involved.

INTERVIEWING TO FIND THE RIGHT CONSERVATOR

When looking for a conservator to assist with your project, determine the role that person needs to fulfill. Will they need experience working with waterlogged materials? Do they need to live nearby or be willing to travel? Is archaeological experience needed or will an objects conservator be suited to the task? Does their laboratory have additional equipment to perform artifact analysis or specialized treatments? It is okay to begin with a generalized idea for how a conservator might be utilized by your project. When interviewing prospective conservators, an experienced candidate will ask questions that will help to develop their role and responsibilities.

During the interview process, ask questions regarding the candidate's background.

- What training/education have they obtained? If they do not have a post-graduate degree in conservation, make sure to ask about their additional experience working under the supervision of senior conservators.
- Are they affiliated with any professional conservation organizations, such as AIC, ICON (Institute of Conservation), IIC (International Institute for Conservation), or others?
- How long have they been in professional practice?
- Do they have a specialization, such as in objects, archaeology, architecture, cemeteries, etc.?

Inquire into their conservation experience.

- Do they have experience working on your specific type of collection/artifacts/site?

- Is conservation their primary work?

- What types of additional services are they able to provide? Material analysis, rehousing, packing for transport, mount manufacture for display/storage?

- Are they in private practice? Government? An academic institution? A museum?

Inquire into their archaeological experience.

- What is their previous involvement on excavations? Ask for specifics: many conservators may have participated in a conservation field school abroad but are unfamiliar with CRM archaeology.

- Ask specifics that pertain to your site. For example, do they have lifting experience for fragile or oversized artifacts? Are they able to treat waterlogged organics?

Address the logistics.

- What is their availability?

- What kind of insurance covers their business?

- How will the collection be transported to and from their facility?

- What is their cost and billing structure?

The individual may have been discovered via a colleague's recommendation, but it is always useful to ask for additional references and examples of previous, and preferably comparable, work.

WHAT TO EXPECT

Upon selecting a conservation professional for your project, the conservator will want to review the scope of work to determine the equipment, materials, and time investment needed. This may include time in the field, travel, and preparatory work. For the treatment of artifacts, the conservator will want to examine the artifacts either in person or using photographs and measurements before proposing a treatment.

As with most specialists, conservators working in the field base their costs on an hourly or daily rate depending on the size of the project. For artifact treatment, conservators should provide a preliminary report that documents the examination of the objects and proposes treatment options with their estimated costs. Depending on the type of project, the treatment proposals may be very specific, especially if the scope is limited to a small number of unique artifacts. Alternately, the proposal may describe general treatment objectives for projects with a large quantity of similar artifact or material types. Conservators should consult with the principal investigator or lab manager if it is later determined that the treatment method needs to deviate from the original proposal.

Most conservators will be working on multiple projects at any given time unless an arrangement has been made for them to work exclusively on your project. This may impact the duration of the project. There is also an element of uncertainty in the treatment of archaeological artifacts. Experience will provide an estimate for how long a treatment may take, but some artifacts or site conditions can present unique challenges, e.g., the desalination of metals is dependent on the chloride content in the burial environment, or concretions may prove more dense than initially evaluated. If there are deadlines or time restrictions, communicate this with the conservator, as they may impact treatment decisions or workflow priorities; but also realize that some processes take time. Conservators should be willing to provide updates on the status of your project on a predetermined schedule or upon request.

Documentation is a cornerstone of the conservation process. The conservator should provide a treatment report for every artifact or artifact group. The length and detail of such reports will vary, but all should include descriptions of the artifacts, details of the treatment performed with materials listed, photography documenting the condition before and after treatment, and recommendations for storage and continued care. A conservation summary document, such as might be included in the final site report, can be requested but is not typical of the documentation recorded in the normal course of a conservation treatment. Treatment reports should be retained with the artifacts, as this information may be of use to future curators, researchers, and conservators in the interpretation and care of these objects.

CONSERVATION PRIORITY AND BENEFIT

Conservators should work together with lab managers, curators, and archaeologists to establish which artifacts are priorities for conservation intervention and will receive the greatest benefit from those actions. This is best accomplished in collaboration, as conservators will approach the selection with a different set of criteria than those who are more familiar with the contextual value of the objects. Conservators generally focus on specific material types that are either more vulnerable to deterioration or have a greater potential for conservation benefit. For example, there is a lot of conservation potential in the reconstruction of a ceramic vessel but no risk of deterioration or loss of potential if reconstruction is *not* performed. By contrast, there is also high conservation potential in the reshaping and freeze-drying of a wet leather shoe but a total loss of that potential, if the leather is subjected to microbiological attack or uncontrolled drying. Although copper alloys are often attractive objects and are comparatively inexpensive to treat, they are generally more stable than, for example, iron that will corrode and spall and potentially lose its diagnostic significance several years after excavation. Observations and generalizations such as these can be made by experienced conservators as they are identifying where to allocate conservation resources for the best value for money. Conservators are also able to assess artifact condition with tools such as x-radiography to help identify artifacts that are physically robust candidates to undergo conservation treatment without minimal damage or loss.

However, the objects selected for treatment by a conservator may not be a priority for the archaeologists, researchers, or stakeholders responsible for funding the project. The context from where the artifacts are recovered have a large part in the selection process. It is a better use of resources to treat objects from significant features or that lend themselves to the interpretation

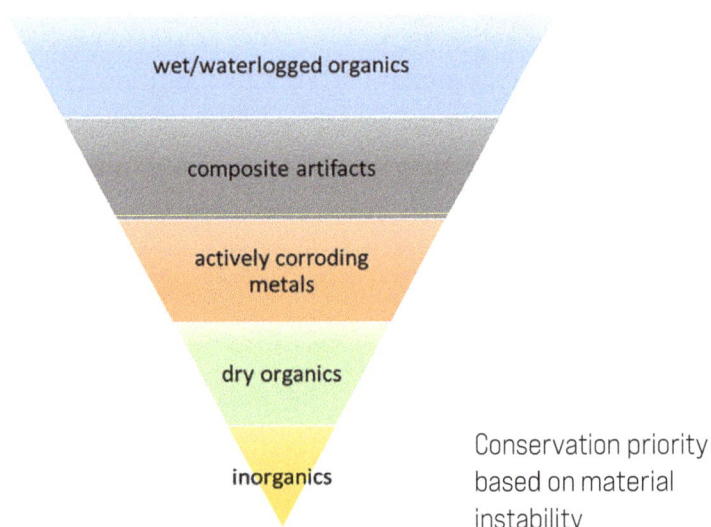

Conservation priority based on material instability

of the site. Stakeholders may select artifacts for display or promotion that are based on aesthetics or that support a particular narrative. Conservators will not have all the necessary information to make those decisions independently. Consolidating the various assessments, analyses, and uses for the collection into a holistic conservation strategy can maximize the preservation value with the resources available.

If an artifact has been deemed so significant as to be included in the site report, either by photo or description, it should be evaluated for conservation so that future researchers accessing the records are also able to access the objects referenced.

Analysis

There are many analytical techniques available to archaeologists to assist with the interpretation of collections. Some of these methods, such as carbon dating and dendrochronology, may assist with chronology. Others focus on materials analysis such as x-ray fluorescence (XRF), x-ray diffraction (XRD), energy dispersive spectroscopy (EDS), mass spectrometry (MS), and many more. Determining what type of analysis is the most suited to answer a particular research question and whether an object is eligible for a certain form of analysis can be overwhelming if one is unfamiliar with the options and techniques. Conservators are well positioned to provide this guidance, advise as to the cost effectiveness of certain tests, explain the sampling process and sizes needed, recommend facilities to conduct the analysis, and assist with the interpretation of the results as it pertains to the object or collection in question.

Some conservators may be able to conduct some analysis in-house where the information facilitates their ability to carry out conservation. Conservators use a range of methods to assist in the identification of particular material types and to assess the condition of the materials being treated.

Some tests are very complex and require specialized chemicals to induce the reactions. Other tests are more rudimentary and can be conducted in most lab settings with experience. All rely on observations to evaluate the results.

Non-destructive tests and observations should always be carried out before embarking on analysis that may cause damage to or loss of an object. When pursuing destructive analysis, take care to minimize the sample size and retrieve the sample from an unobtrusive location.

OPTICAL MICROSCOPY

An optical microscope is an essential tool in any conservation lab. Viewing materials under high magnification can often reveal identifying features that are difficult to observe with the naked eye.

UV-INDUCED VISIBLE FLUORESCENCE

The human eye cannot normally see UV light, which travels in short wavelengths outside our visible spectrum. However, certain materials absorb UV light and reflect it back as visible light. This fluorescence can be a contributing identifier for certain objects, including ivory, tortoise shell, horn, certain pigments, and adhesives.

ODOR CHARACTERIZATION

The surface molecules of an object are excited when heat is applied, producing a smell that can be evaluated by a sensitive nose. For some materials, friction (rubbing an object between one's hands) may be enough to induce an odor. Other materials require a higher temperature. This can be achieved by heating a pin or needle over a flame and touching the tip to a small, unobtrusive area on the object. Certain material types may scorch or melt, so take care with the area selected and record the results to minimize the need for additional testing.

X-RADIOGRAPHY

The use of x-radiography in post-processing collection is unfortunately infrequent. It has been available for over one hundred years, yet it is underutilized in archaeology, often because it is wrongly perceived as expensive or inaccessible. However, x-radiography is both affordable and a very useful tool. The documentation produced by x-radiography can more accurately catalog a collection, be part of a sampling and discard strategy, assess artifact condition, and assist in determining conservation priorities. And it is nondestructive.

X-radiography is commonly used to assist in the identification of concreted metal objects. It can also display the original surface dimensions of an object for more accurate measurements. It is possible to use x-ray images to identify different material types based on the density and corrosion characteristics present in the image and to reveal surface details and markings that are obscured by corrosion products. These features lend themselves to more accurate artifact catalogues and interpretation.

When used to survey bulk corroded metals, such as nails, x-radiography can be part of a responsible sampling and discard strategy for collections from historic sites. Its use can document materials more accurately than photographs or measurements of the corroded objects, save space, decrease curation fees, and offer better artifact identification.

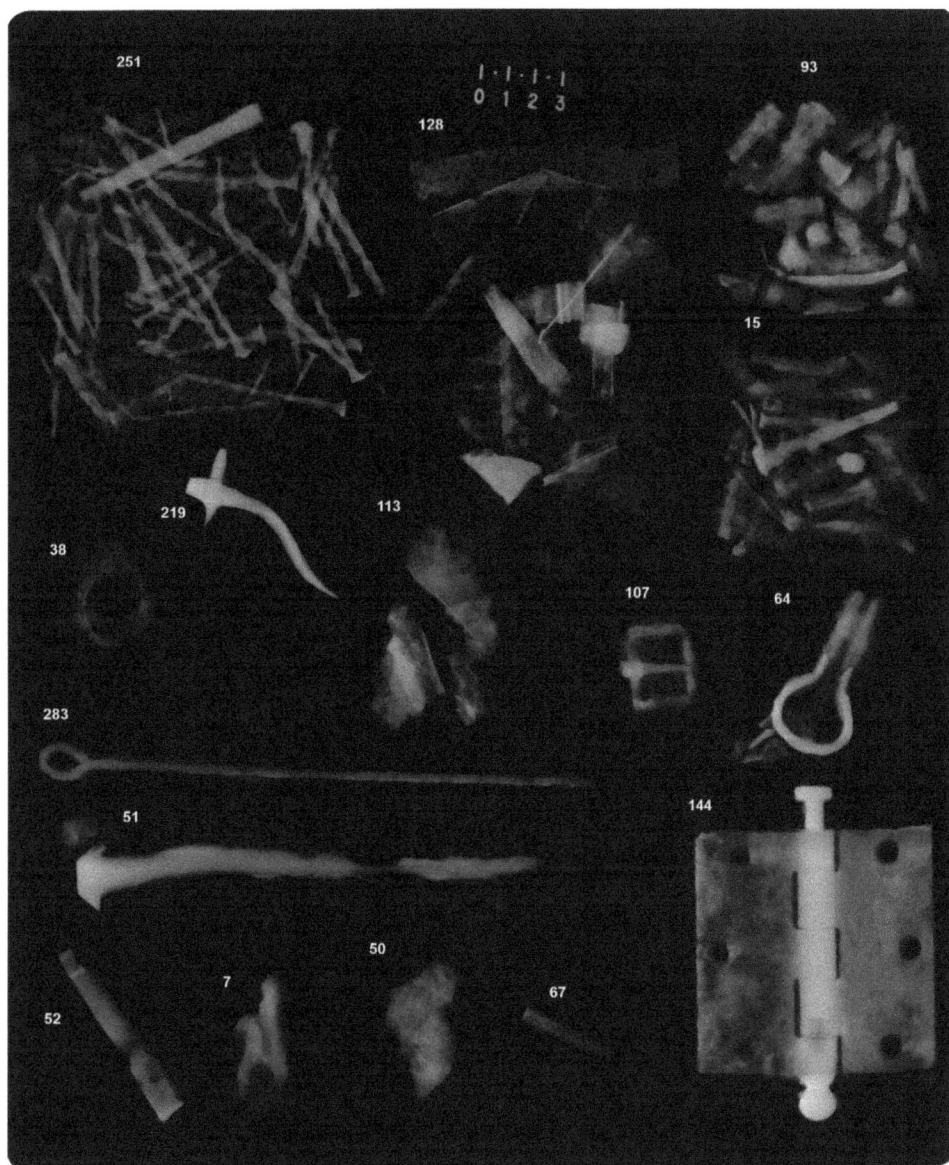

X-ray image of iron assemblage. Artifacts can be identified and measured; darker areas show areas of poor preservation or replacement of the metal by corrosion.

X-radiography can also be used to assess the condition of artifacts by observing the quantity of metal preserved among the corrosion products. This is especially useful to conservators and collections managers to help prioritize artifacts selected for treatments. X-ray images can determine if an object is a viable candidate for conservation, thus allowing conservators to best utilize the funds available for treatment.

While it is very useful to have material x-rayed by technicians experienced in archaeological collections, this service may not be available locally. Instead, it is possible to collaborate with other professional groups who utilize x-radiography. Universities, hospitals, large animal vets, dentists, even a technician manning the x-ray scanner at the local courthouse or airport may be willing to assist in examining your collections. These images can then be shared with colleagues to assist in their interpretation.

Collaborations

Archaeology is a collaborative field to its core. The study of human history and its marks upon material culture and landscape is nuanced and variable, requiring the expertise of researchers, osteoarchaeologists, archaeobotanists, geographic information system (GIS) specialists, historians, and many others to compile the body of information that allows archaeologists to interpret a site. And the clues left behind can be so very fragile and fleeting. The decisions made during recovery and with the assistance of conservation professionals will determine how much of this information is preserved for the access and use of our colleagues. Conservation is not exclusively concerned with the treatment of artifacts and is not based on some recipe book. The guiding principle of archaeological conservation is to facilitate the use of archaeological resources for interpretation, for analysis, for research, for professional colleagues, for the public, for today and for the future.

WORKING WITH CONSERVATORS

Kerry Gonzalez, Senior Principal Investigator,
Gray & Pape Heritage Management

INTRODUCTION

Archaeologists often find themselves on the other end of a call with a conservator at some point in their career, especially if they are either leading excavations or have a managerial role in a lab. The importance in establishing a relationship with conservators cannot be overstated, and fostering such relationships can only benefit your individual projects and the archaeological record in the end.

Throughout my tenure as a lab manager for a CRM firm in the mid-Atlantic, I was able to work, on a very deep level, with conservators and collections managers alike, to provide advice, guidance, and information on best practices regarding the collections I was managing. Throughout this evolving relationship, I began to look at collections in a different light and became better at my job.

There are unfortunate misconceptions about the use of conservators, especially in a CRM environment. These environments are largely controlled by the need to stay on budget within a specific time frame, and we are all aware that conservation of artifacts costs extra money and can take time. However, it is a goal of this article to highlight how these extra costs can be mitigated by working with a conservator. If you are able to build the extra funds into your budgets, the outcome will be even more beneficial to the project and to the collection associated with the site(s) you are investigating.

CASE STUDIES

In an effort to highlight how this relationship can benefit a project, the public, and the overall assemblage, the following section provides two case studies on two unique historic sites. The first project discussed here includes the Phase I–III assemblage; thanks to the generosity of the Delaware Department of Transportation and the Federal Highway Administration, there were sufficient funds for conservation. The second project highlighted is a large Phase III on a historic site where conservation was not included in the original budget; however, some cost-saving measures allowed the archaeologists to go the extra mile and work with conservators through x-radiography of certain iron objects (González and Salvato 2019).

Houston-LeCompt

In 2012 the Houston-LeCompt site, a late-18th through early-20th century domestic site in New Castle, Delaware, was excavated at the Phase III level. The Phase I–III assemblage for this site contained over 50,000 artifacts, many of which required conservation measures. While typical conservation on the recovered iron and copper alloys was completed, the material that required the most 'heavy lifting' were wooden well cribbing fragments. At the time of recovery it was unknown if formal conservation would be completed on all, some, or none of the artifacts; space and funding were being taken into consideration. As these pieces needed to remain stable during a nine-month waiting period, the archaeologists caring for the well timbers were in constant contact with the conservators at the MAC Lab. Not only were the conservators instrumental in providing a list of chemicals to treat the timbers should any mold develop, but they also helped develop a strategy for keeping the wood submerged in a small area that would also permit easy water-changing efforts. When the time came to transport the materials to be professionally conserved, conservators were once again there to offer assistance on the best way to stabilize the materials for the two-hour trip.

Beyond their assistance with the well cribbing fragments, from time of recovery to finalized treatment, conservators also helped make conservation

Sample of well cribbing recovered from the Houston-LeCompt site. (Photo courtesy of DelDOT and the Federal Highway Administration)

decisions on the other materials in the collections, such as buttons, a broad hoe, a brass spigot, and coins. These tough choices were a complete team effort that began with the important task of x-raying to (1) help identify diagnostic or interpretive objects masked by years of corrosion; (2) reveal how much metal was left in an object, thus possibly negating the need for conservation if only a mass of corrosion existed; and (3) help identify the material composition.

Once this important data has been obtained from x-rays, discussions between the archaeologists and conservators can begin. Many factors should be taken into account, including budget, timeframe, stability of the object, significance of the artifact to the overall collection, and interpretive potential. Because there are so many factors involved, these decisions should never be made alone. Communicating with a conservator is essential.

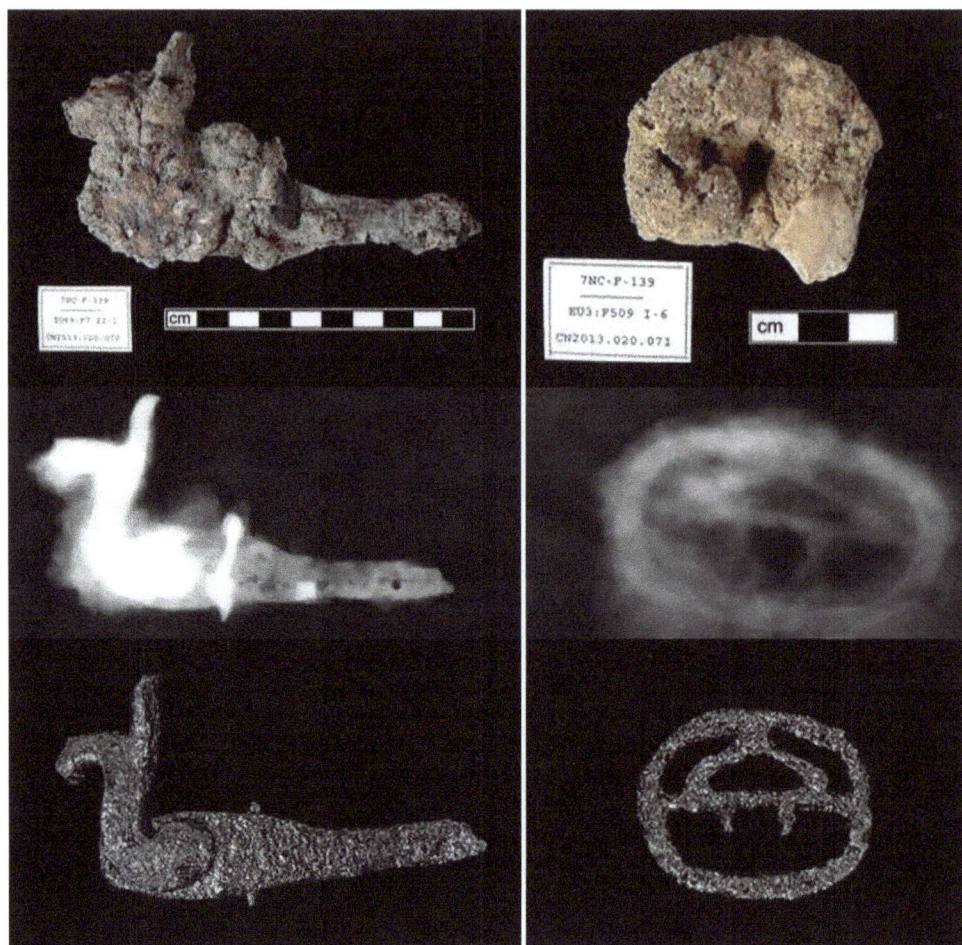

Selected artifacts from the Houston-LeCompt site examined through x-radiography: a percussion lock (left) and shoe buckle chape (right). Image shows artifacts before treatment, x-rays, and after conservation. (Photo courtesy of DelDOT and the Federal Highway Administration)

Squirrel Creek Site

Like the Houston-LeCompt site, the Squirrel Creek site was also excavated at the data recovery level and produced over 22,000 artifacts along with a wealth of data. However, unlike the Houston-LeCompt site, no artifacts were conserved as part of this effort. Objects from the collection were only x-rayed to aid with identification, including of a few objects' manufacture methods. The excavation and ensuing lab work is an excellent case of having to prioritize funds, specifically lab monies, in order to gather the best data to aid in the interpretation of the site.

Once again, archaeologists worked with conservators on this approach and thus were able to determine that many of the highly corroded indeterminate nails, thought to be framing or building nails, were actually horse and mule shoe nails. Not only that, but they were also able to determine the manufacture method and repair of a recovered iron frog gig. This artifact was either hastily manufactured or modified, as evidenced by the x-ray image showing the areas where the gig was heat-joined. One interpretation

Iron frog gig: x-ray (left) and before treatment (right). (Photo courtesy of the North Carolina Department of Transportation)

of this manufacture style is that the gig was repaired after one of the original three tines had broken (Hatch et al. 2017).

While no artifacts from this collection were formally conserved, a select few were assessed for stability by the conservators, who also utilized their skills during the x-ray of the iron objects from the collection. This approach is a great example of the diverse services available to archaeologists who work with conservators and build a collegial relationship that benefits both parties.

BEST PRACTICE

When collaborating with any group or individual, the two most important features of a successful project are communication and respect. It is important to know that conservators have a unique skill set that should be utilized to the fullest extent possible. However, it is also very important for them to know the established research goals of the project and what you hope to gain from working with them. Are you simply looking to stabilize some fragile objects? Are you hoping to learn more about the collection by employing x-radiography to aid in artifact identification? Are any of the objects slated to be put in a temporary or permanent exhibit? These are all factors that should be discussed at the beginning of your conversation with conservators. Once they have this information, they will better understand your goals and objectives and can make recommendations accordingly.

Equally as important is to consider consulting with a conservator during the proposal phase of a project. While it is difficult to determine the quantity of objects that may need conservation or x-ray during this phase of a project, simply having a line item in your budget for a set amount is a good first step. If your research questions/goals are more artifact/material culture-based, then you can easily justify increasing this budget amount. Additionally, having a set amount of funds slated for conservation efforts can help a conservator make recommendations on what items can be conserved. For instance, large iron objects cost more, and you may choose to conserve a large number of smaller copper alloy artifacts instead.

SUMMARY

The goal of this article has been to highlight the importance of working with conservators on your projects, ideally at the proposal phase. There will be situations where unexpected finds will result in frantic phone calls to a conservator, but that should be the exception and not the rule. Establish a relationship with the conservators in your area or where your project is taking place. Ask for their advice and listen to it.

González, Kerry S. and Michelle Salvato. Pictures Speak for Themselves: Case Studies Proving the Significance and Affordability of X-ray for Archaeological Collections. In *New Life for Archaeological Collections*. Rebecca Allen and Ben Ford, eds. Lincoln: University of Nebraska Press, 2019.

Hatch, D. Brad, Caitlin Sylvester, Kerry S. González, Michelle Salvato, and Mike Klein. *Archaeological Data Recovery at the Trogdon-Squirrel Creek Site, Randolph County, North Carolina*. Fredericksburg, VA: Dovetail Cultural Resource Group, 2017.

PUBLIC OUTREACH ARCHAEOLOGY AND ARTIFACT

Aaron Levinthal, Senior Archaeologist,
Maryland State Highway Administration

INTRODUCTION

Since 2006 the Maryland Department of Transportation (MDOT) has funded public archaeology projects focused on African American history along the state's transportation system. The program is led by MDOT's Chief of Cultural Resources and supported by a team of consultant archaeologists and specialists. Each project is approached with a robust research agenda and is benefited by historical research, excavations, laboratory analysis, artifact conservation and curation, specialized analysis (e.g., DNA, macrobotanical, and faunal analysis), and a series of reports. The capstone of each project includes various interpretive products, site tours, public presentations, and special events for the projects' partners and any community descendants.

Each MDOT project has included partnerships with federal, state, and/ or local agencies, private property owners, and descendants. Often, these partners are unfamiliar with archaeological methods and may be skeptical of the science. Some community members may even hold a negative view of archaeologists who they see as obstructionists to development and land rights. Most partners, however, are elated to have their history chosen to be studied by professional archaeologists, knowing the process can result in new discoveries and unexpected data. Every project has demonstrated how science can successfully incorporate a variety of stakeholders from diverse backgrounds and insert them into the archaeological process. We have been able to intelligently gather data to be synthesized and distilled into meaningful products digestible by all audiences.

Each MDOT archaeology stewardship project has produced important narratives, but even the most well-presented and engaging history can still result in dismissal and loss of interest by an audience. To some audiences, stories of a historic event, person, or place will not be impactful unless it is accompanied by material culture such as artifacts recovered from an archaeological site.

Indeed, the most important goals of any public archaeology project are the encouragement of conversation and cooperation, and the appreciation of the partners' contributions to the site. Most MDOT projects are intensive studies, and cutting-edge techniques are commonly applied to guide projects towards important results.

THE VALUE OF CONSERVATION

In order to ensure that the undertaken research is shared and available into the future, MDOT outreach projects include conservation. The most used services are artifact conservation and stabilization and x-radiography of metal and composite artifacts.

Many artifacts, such as those made of bone and ceramic materials, come out of the ground stable and do not need further stabilization by a conservator. Other artifacts, however, quickly corrode and break down as soon as they are pulled from their sub-surface micro-environments. Ferrous objects are infamous for appearing amorphous once they have been excavated from site soils. Conservators can efficiently identify an artifact hidden within a mass with x-ray technology. Other artifacts can be readily identifiable and data-rich in the field but may be extremely fragile and in need of immediate stabilization and conservation in order to preserve interpretive value after excavation. A metal button or coin, for example, is an artifact type that may have designs or dates that are not easily visible when removed from a site, but a conservator can stabilize the object and even facilitate the reading of the text on a button back stamp, or the year and other details on a coin.

During and after fieldwork, artifacts are selected to help tell and illustrate the story being uncovered by the archaeologists. These artifacts are visual aids, and the physical evidence from the past is extremely effective in engaging with the public. Artifacts associated with activities (e.g., buttons, dishes, jewelry, toys, etc.), are extremely effective and help explain behaviors that were part of the daily lives of people in our past.

An archaeologist discusses a selection of artifacts recovered during the investigation of enslaved family quarters in Anne Arundel County, Maryland with project partners and members of the descendant community. (Photo courtesy of the Maryland Department of Transportation)

Since MDOT partners with various stakeholders on several public outreach projects each year, and because each project includes at least one interpretive event, their archaeology laboratory coordinates with conservators frequently. Many project staff have attended and completed conservator-led workshops on artifact identification and field conservation strategies for archaeologists. The small group, hands-on training includes introductions to cutting-edge conservation techniques, materials, and technologies as well as lessons on best-practice methods for handling a variety of types of unanticipated field discoveries and artifact recoveries. The workshops provide opportunities for detailed question and answer sessions, and the resulting information updates MDOT field and lab methodology and practice.

All archaeological projects should incorporate artifact conservation into their scope of work to ensure the stabilization of collections. Without conservation of an artifact assemblage, the current and future research value of an archaeological project is limited and compromised.

CONSERVATION TRAINING AND CONSULTATION

Elizabeth Waters Johnson, Laboratory Manager/Principal Archaeologist, Wetland Studies and Solutions, Inc.

INTRODUCTION

My introduction to conservation came in graduate school in a course taught by a well-regarded conservator at the Smithsonian Museum of Natural History. The introductory course entailed a great deal of chemistry and a primer on what had gone dreadfully wrong with conservation efforts in the past. Most of the discussions focused on conservation in a museum setting, so when I began working in cultural resource management (CRM) it was difficult to translate the lessons I had learned in this class to the work I now found myself doing. It wasn't until I began working on a large-scale urban project with artifacts from diverse contexts that complications involving conservation began to arise. With the exciting discovery of an 18th-century ship hull along the waterfront, my firm contacted a professional archaeological conservator to guide us through the necessary steps to preserve the waterlogged timbers and keep them from deteriorating any further until they could be moved to a temporary storage location. As I watched the crew apply this advice in the field I wondered what else we were recovering that might need additional care. Lack of communication meant that we were often finding surprises in the laboratory that should have been brought to our attention immediately when the artifacts were recovered in the field, and despite being surrounded by well-respecteded and experienced archaeologists, there didn't seem to be a lot of knowledge regarding field conservation methods.

The trial and error experienced with this urban project left me extremely frustated and determined to do better moving forward. The guidence we received from conservators was invaluable, so when they announced that they would be holding a workshop on field conservation methodology, I was quick to register.

The workshop provided hands-on and realistic field methods, encouraged us to cultivate relationships with our local conservators, and emphasized the need for communication between not just the laboratory and the field, but between all stakeholders. The knowledge gained from this workshop, and from others that followed, allowed us to create a methodology that improved collaboration between the laboratory and field crews, enhanced the work we provided our clients, and even enriched our contribution to the historical record.

THE FIELD

The crucial first step in the process was to improve the communication between the laboratory and field crews. Armed with knowledge from the workshop, we created a document of quick field conservation methods, which we had laminated and placed in the field dig kits to serve as a guide for what to do with artifacts like leather, bone, textiles, and wood found in either dry or water-logged contexts. We also provided some basic materials in the dig kits—paper towels, plastic bags, lysol, 4 mil thick black plastic sheeting, and zip ties—that the crew might need to perform these quick field conservation methods. We then met with the field crew to review the

18th-century ship hull recovered from 44AX229. (Photo courtesy of Thunderbird Archaeology, a Division of Wetland Studies and Solutions, Inc., a Davey Tree Company)

information and answer any questions they had. We discussed the kinds of artifacts that might need additional care and encouraged the crew to contact us with any questions.

One of the most memorable and most useful portions of the conservation workshop was the discussion dedicated to "MacGyvering" artifacts that you recover in the field. Yes, the name—referring to making or repairing something with the materials you have at hand – is entertaining and conjures images of a scrappy protagonist who uses his intellect along with gum and duct tape to fight crime, but it was also the most realistic application of conservation knowledge to a field crew working on an active archeological site. Artifacts recovered from diverse contexts needing conservation are often found in an urban setting, where the pace of field work is, shall we say, expeditious. It is a balancing act in which we as professional archaeologists strive to recover the most accurate data as quickly as possible, usually in the least ideal circumstances imaginable. Knowing, at the most basic level, what we need to do to keep an artifact intact and what we can use to do that is extremely advantageous in such circumstances.

As the laboratory director I can plan, in great detail, the dig kits that will be provided for field conservation of recovered artifacts, but it is up to those actually on site to know how and when to use those items. When combined with training and reference guides, MacGyvering allows the field crew the flexibility to make good choices and apply their knowledge with the materials they have on hand. And truth be told, while I can provide all of the necessary equipment the crew needs to wrap the artifacts for transport back to the laboratory, they are going to run out of items or misplace them or forget that they even have them, so providing the crew with the tools to be flexible and the knowledge to use anything they have on site to temporarily stabilize the items until they reach the lab is essential. This allows them to recognize a situation as basic as, "I need to keep this wet but I don't have what we would normally use. What can I use to keep this wet until we get back to the laboratory?" Providing the crew with alternatives, or MacGyvering, is just as important as providing them with textbook best practices.

It isn't enough to just impart this knowledge to the crew and hope that they make good choices. Communication also plays an incredibly important role in the conservation of artifacts recovered from the field. What good does it do if the field crew bags a piece of leather with damp paper towels, places it in a plastic bag, and then tosses it in with the hundreds of artifacts piled in the laboratory if they do not mention what they've done to the laboratory staff? The artifact might not be found for weeks, at which point it might not be as well-preserved as we would like. It is important to check-in with the crew regularly and make sure they contact laboratory staff when they find something that requires special treatment when it arrives at the laboratory.

We implemented a strategy to ensure sensitve artifacts in need of conservation arrived at the laboratory from the field in a timely manner. For example, if wet leather was recovered in the field, it would need to be bagged with wet paper towels and placed in a second, clearly-labeled plastic bag that had been sprayed with lysol to prevent mold. Next all the wet artifacts would be placed together to be returned to the office as soon as possible. The crew would notify the laboratory that sensitive artifacts had been recovered and specify when they would be brought to the office. Once the artifacts arrived at the office they would be registered on a specially designated list and placed in the refrigerator. This would eliminate the chances of finding surprise, dried-out leather buried in artifact bags that had been turned into the laboratory weeks prior. The laborary then maintained a schedule for examining artifacts stored in the refrigerator. Originally we decided to do spot checks on a biweekly basis; however, over time we deemed spot checks once a month to be adequate. Once the artifacts were in the laboratory, cleaned and properly stored, laboratory staff could discuss next steps with the conservators.

CONSERVATION CONSULTATION

Having implemented field conservation procedures, consultation with our local conservator usually begins on a project once sensitive artifacts are recovered. This typically entails sending a lot of quick and dirty photos and requesting expert advice on how best to store the artifacts as well as whether or not conservation is deemed necessary. The conservators help us prioritize which artifacts are in need of conservation versus those that might benefit should the budget allow. They also provide proposals for treatment costs that we can use to help guide our clients. Communication with the client is ongoing regarding the types of artifacts that might need conservation and the costs required for such treatment.

It may not always be easy to discern which artifacts need conservation. A conservator can provide guidance as to which artifacts appear stable and

Before and after treatment images of a silver knee buckle with iron tang recovered from 44FX2429 (Photo courtesy of the Maryland Archaeological Conservation Laboratory).

require less conservation than those that would benefit from treatment. For example, this men's knee buckle recovered from site 44FX2429 was comprised of two different types of metal. The ferrous metal had corroded, while the silver portion of the knee buckle appeared stable. However, the conservator pointed out that the interaction between the two metals was likely to cause further deterioration over time; thus conservation was prioritized.

X-radiography can also help us determine whether conservation is needed. It allows us to see how much of the metal remains in the artifact as well as if the corrosion is obscuring artifacts that could be revealed through conservation treatments, such as this intaglio recovered from site 44AX0235. While we could see a portion of the chalcedony face, it wasn't until the artifact was x-rayed that we were able to see a portion of the clasp hidden by the deteriorating metal. Although we already knew conservation was necessary for this artifact, the x-ray provided the conservator with more accurate data on which to design the treatment strategy.

Chalcedony and copper alloy intaglio recovered from 44AX235. (Photo courtesy of Thunderbird Archaeology, a Division of Wetland Studies and Solutions, Inc., a Davey Tree Company)

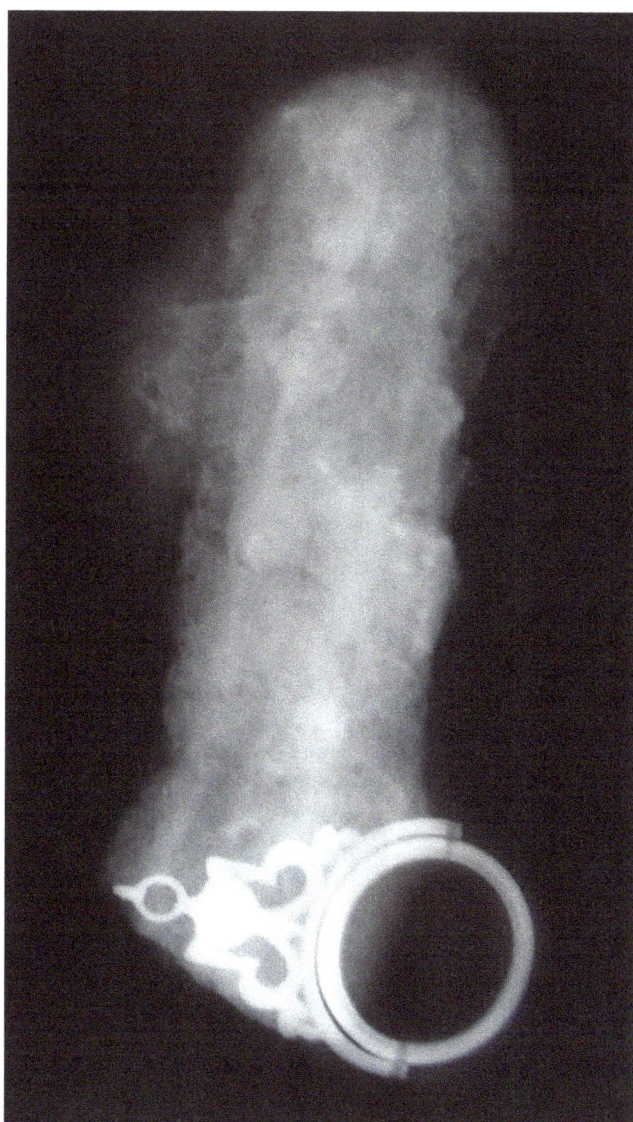

REPOSITORY CONSULTATION

It is always best to consult with the final repository where the artifacts will be curated prior to sending the artifacts for conservation. Including the repository in the process early will allow the conservators to provide treatment plans that are consistent with the repository's requirements as well as address any concerns that the repository might have. The repository usually has its own priorities and limitations regarding conservation of their collections. While most would prefer that all sensitive or unstable artifacts be conserved, some repositories are limited in their ability to stabilize particularly problematic artifacts, or have limited space for artifacts that require specialized storage. It is also imperative to provide all of the conservation documentation to the final repository for their records. Involving the final repository early in the process means a much more efficient outcome for everyone involved.

CONCLUSION

Every project in CRM is unique and provides new opportunities and challenges. The adventure is that you never know what you are going to find —nevertheless, you had better be prepared. While encountering artifacts in need of conservation may not happen on a daily basis, it is important to recognize when you have recovered something that requires special care. While there will always be surprises along the way in CRM, having a plan in place when they do arise and resources to consult allows you to work artifact conservation into your workflow, keeping the project on time and on budget. Thanks to field conservation training provided by professional archaeological conservators, we were able to create a process that cultivated better collaboration between the field and the laboratory crews as well as with our local conservators and the final curation repositories, which has ultimately strengthened our contribution to the historical record.

Further Reading

Artifact preservation is complicated and multifaceted. Material type, manufacturing techniques, state of preservation, burial environment, and preservation resources all play into the decision-making process. The conservation advice in northern Europe will differ from advice on Mediterranean sites, which will differ from post-contact North American advice. But the science and methodologies remain the same. Conservation, as with archaeology, has been performed at the avocational level, but experts in our fields make it a lifelong study built upon many years of experience and professional development. Both conservators and archaeologists would benefit from an expansion in crossover resources.

Appelbaum, Barbara. *Conservation Treatment Methodology*. Routledge. 2007.

Cronyn, J. M. *The Elements of Archaeological Conservation*. Psychology Press. 1990.

Pedelì, Corrado, and Stefano Pulga. *Conservation Practices on Archaeological Excavations: Principles and Methods*. Getty Publications. 2014.

Sease, Catherine. *A Conservation Manual for the Field Archaeologist*. Cotsen Institute of Archaeology. 1994.

Watkinson, David, and Virginia Neal. *First Aid for Finds*. 1998.

Weaver, Graham. *An Introduction to Materials*. Psychology Press. 1992.

Acknowledgements

I would like to express my deepest appreciation to the Foundation for Advancement in Conservation (FAIC) and the Samuel H. Kress Foundation for providing the financial support to enable this project, and the Society for Historical Archaeology for their publication support.

My sincere thanks to my colleagues at the Maryland Archaeological Conservation Laboratory and the Virginia Department of Historic Resources for reviewing manuscripts and pulling artifacts for photography, and to Tom Kutys, Kerry Gonzalez, Aaron Levinthal, and Elizabeth Waters Johnson for their thoughtful contributions.

I am grateful to my family and friends for giving me the time and space to put these words onto paper, it would not have been possible without your support and generosity.

www.ingramcontent.com/pod-product-compliance
Lightning Source LLC
Chambersburg PA
CBHW042339030426
42335CB00030B/3402